The Cambrian Explosion

The Cambrian Explosion

Evolution's Big Bang?
Or Darwin's Dilemma

An engineering professor gives expert-witness testimony
on the origin of animals, including many novel analyses.

Walter Starkey

COLUMBUS, OHIO
2012

To order additional copies of this book, contact:
Xlibris Corporation
1-888-795-4274
www.Xlibris.com
Orders@Xlibris.com
100448

PREFACE

Every book must, to some extent, reflect the experiences and imaginations of its author. This is particularly true in the case of this book. I have been interested in animals all of my life, including the matter of origins. I have studied the theory of evolution and the theory of creation since I was 16 years old. Entering college, I chose mechanical engineering as my profession, and, after earning the BME, MSc, and PhD degrees in mechanical engineering, I became a professor of mechanical engineering, which occupied two-thirds of my professional career. The other one-third of my career was devoted to serving as an expert-witness in law cases in which engineering matters were involved. All of these experiences have come together as I have written this book. The book involves engineering analyses of animals and the early origins of the earth. The format of the book consists of a court trial in which I, as an expert-witness, have accepted the assignment of performing engineering analyses to determine, with engineering certainty, who designed and constructed the animals of the earth, and who caused the Cambrian Explosion.

During the past five years I have devoted all of my professional time to the project of doing research for, and writing, this book. The book has been meticulously researched and composed. I often have spent a half-hour deciding on a single word. Although this book should be of interest to scholars and intellectuals, I have written it primarily for young people at a high-school or college level of understanding. This is not a religious book. Its conclusions have not been arrived at on the basis of faith and doctrine. The conclusions of the book have been based on scientific facts and engineering analyses.

If you are an atheist, by all means, read this book, from cover to cover, and give copies of it to your friends. On the subject of chronology, this book has been written from the perspective of old-earth scientists. If you are a young-earth creationist, don't fret, and don't summarily reject this book, just assign your own dates to the events mentioned in the book. The cardinal message of this book has to do with *who* created the animals, and it is not important *when* they were created. And, if the scientists of the world eventually come to the conclusion that the earth is 10,000 years old, I want to be the first to congratulate my esteemed colleagues who are young-earth creationists.

As you will find out when you read this book, I did in fact determine who caused the Cambrian Explosion. This book offers sound answers to many interesting questions. It contains several new and original explanations for observed facts and phenomena, which explanations have never before been suggested. I think you will find that the reading of this book will be an exciting adventure. I hope you will enjoy it.

Walter L. Starkey

CONTENTS

THE CAMBRIAN EXPLOSION
EVOLUTION'S BIG-BANG?
OR DARWIN'S DILEMMA?

AN ENGINEERING PROFESSOR GIVES EXPERT-WITNESS TESTIMONY ON THE ORIGIN OF ANIMALS, INCLUDING MANY NOVEL ANALYSES

CHAPTER 1. INTRODUCTION

Spider

Rock

Could You Make a Spider?

How long would it take you to make a spider out of a rock? That is, I think, a profound question. So I will ask it again for emphasis. How long would it take you to make a spider out of a rock? Think about it before you answer. It might take quite a while, because a spider is really a very complex creature. But you are very smart. Human beings are much smarter than all of the other animals on the earth. You can invent, design, and construct complex things. But are you smart enough to construct a spider? If you are inclined to admit that you might not be up to the task, think again. Maybe you could do it if you worked at it for a long time, like 10 years, 100 years, or a million years, if you could live that long. I don't think you could ever do it! Some things are just too difficult for us to do, regardless of how long we might work at it.

Why do I ask you to make a spider out of a rock? That's because, on this earth, before there were any plants or animals or soil, all we had was rocks. Our earth was just a big rock, with little rocks of various sizes on the surface. But today we have spiders; so, somehow, a rock turned into a spider.

I'll help you in this project by giving you two other ingredients. You can also have a little water and a little air. Our primitive earth did have all three, rocks, water, and air. So maybe you could make a spider out of a rock, with a little bit of water and some air. But even with these added ingredients, I don't think you are smart enough ever to complete the construction of a spider. It's beyond your capability!

Now here is another question. What do you think is the likelihood that a rock could turn into a spider by itself, without any help from you, or anyone else? That achievement would be vastly more difficult than for you to do the job. If you can't make a spider, how could a rock possibly turn into a spider all by itself? If that could be done it would certainly be the most astounding magical trick I ever saw. I'd like to see a rock turn into a spider.

Do you think there are people on earth who believe that a rock can turn into a spider? Would such a person have to be a lunatic? In fact there are people who believe this. I'll tell you who these people are. Almost all of the biologists in the world believe it. And, since they are the experts in the field of spiders and biology, they have succeeded in convincing most of the rest of the people in the world that this can be done and has been done. The people of the world just take their word for it, and they believe it too. It's taught in the schools, and the students believe it. What we're talking about here is the theory of evolution.

Theories of Evolution and Creation

There are two major theories which attempt to explain how the animals of the earth came into existence. They are: (1) the theory of evolution, and (2) the theory of creation. The theory of evolution asserts that all living things evolved by themselves, as a result of the natural actions of the chemical, electrical, mechanical, and other such forces of the universe, without any assistance from any designer-craftsman, or anyone else. Those who believe in the theory of evolution will be called evolutionists. The theory of creation asserts that all living things were designed and constructed by some highly-skilled designer-craftsman. Those who believe in this theory will be called creationists.

2

Purpose of this Book

The purpose of this book is to examine, from the viewpoint of an engineer, the evidence in support of one or the other of these two theories, and, if it should be determined that the animals were created by some supernatural designer-craftsman, it will be our objective to discover just who was this expert designer. I, the author of this book, am an engineer, a professor, and an expert witness in law cases. My specialty is machinery, machine design, the design of machinery. In this book I will show that animals are machines, and I will show that, just as machines do not come into existence without a designer and a builder, neither could animals come into existence by themselves, with no designer and no builder.

Since I have had considerable experience serving as an expert witness in law cases, and in testifying in court trials, I will use this law-suit format within which to develop the material contents of this book. I will use my expertise as an engineer to determine and analyze facts, and my experience as an engineering-investigator in law suits to develop engineering-based opinions as an expert witness. In particular, I will attempt to determine the true origin of the animals of the earth. I will limit my studies to animals, more than the other forms of life on earth, because animals are really *machines* that have been designed, and they are clearly subject to in-depth engineering analysis by someone who is qualified as an expert in the design of machinery.

Interesting Question Answered in this Book.

This book contains many new ideas and novel analyses never before presented to any audience. To give you a brief preview glimpse into what is ahead as you read this book, I have prepared a list of interesting questions that are answered in this book. The list follows, and, for each question, the Chapter is identified which contains the answer.

INTERESTING QUESTIONS ANSWERED IN THIS BOOK:

QUESTION   SEE ANSWER IN CHAPTER

1. Could you make a spider out of a rock?   1
2. Who were the "Ape Men?"   22
3. Is the Neanderthal Man mentioned in the Bible?   22
4. What did God and Henry Ford have in common?   22
5. How did the insects learn advanced vibration theory?   12

6. Did God intermittently visit the early earth to create animals? 22
7. Is the Big Bang mentioned in the Bible?   21
8. Can you explain how the human eye could have evolved?   15
9. Can mutations really produce new species?   15
10. Why aren't machines filled with nerves and blood vessels?   7

11. Can "intermittent visits" fully explain all of the fossils?   22
12. Why don't animals have gears, shafts and V-belts?   7
13. Can new species really arise without changing the DNA?   15
14. What is the true role of natural selection?   18
15. Do the fossils really support the theory of evolution?   22

16. Can "steam-table engineering" explain the early earth?   20
17. Could a water-moccasin snake kill a man on a deserted island? 24
18. Did God regularly make prototype animals?   22
19. What is the greatest scientific mistake of all time?   25
20. Are animals really machines?   6

21. What explosion occurred 540,000,000 years ago?   25
22. What is the role of an NFPA in a theory of evolution?   13
23. Was this book written by an engineer, a scientist, or a professor? 33
24. Can you explain how a bird can fly directly upward?   12
25. How can a man survive in a speeding car that has no brakes?   24

26. Do protozoa need DNA?   10
27. Are the web spinnerets of spiders worthy of being patented?   11

CHAPTER 2. WHAT IS AN EXPERT WITNESS?

Woodchipper Explodes

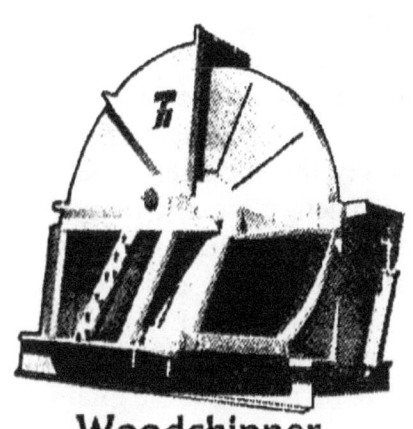

Woodchipper

On July 30, 1974 a woodsman was operating a huge woodchipper in a clearing in a woods near Caldwell, Ohio. A woodchipping machine transforms huge logs into wood chips. The chips are used to make paper. This machine has within it a heavy steel disk about six inches thick and six feet in diameter. The disk has cutters and cutter hold-down bars on it, and it rotates at high speed when the machine is operating. This is indeed an awesome machine and it operates with a deafening roar. On July 30, 1974, all of a sudden, the machine exploded. Parts flew in all directions. It was a centrifugal explosion. One part, a 100-pound piece of steel, a cutting-blade hold-down bar, hit the operator in the chest. He no doubt died instantly. There were no witnesses to the accident. I repeat, there were no witnesses to this explosion!

Within hours after the accident occurred, representatives of the company which insured the machine retained me to go with them to the site. We left immediately. My job was to study the demolished machine, study all of its widely scattered parts, take measurements and photographs, and determine what happened. I was asked to find out why it exploded and who was responsible for the explosion. Even though there were no witnesses, it was my job to perform engineering studies and determine what happened and who caused it to happen.

After many more trips to the accident scene, and after many hours of engineering study, I determined that one of the main bearings in the machine was poorly designed and it had failed. The shaft then moved slightly and a cutter blade hit the mating shear ledge. The cutter and its hold-down bar then tore lose from the disk, and the bar flew at high velocity, hitting the operator. Many other parts of the machine then centrifugally exploded. Who caused the accident? The designer of the bearing caused the explosion!

This accident then developed into a law case of wrongful death. The widow of the deceased woodsman brought a lawsuit against the manufacturer of the woodchipper, and the designer of the bearing that failed. My role in this case, as in all of the many law cases with which I have been associated, was to study the engineering facts of the case, and determine what happened. Then, typically, I am asked to write a report on my findings, or verbally explain my findings before a group of lawyers in a deposition. When the case comes to trial, the judge must first determine if I am really qualified to be an expert witness. The judge studies my qualifications. If he decides that I am qualified, I then testify in court. I report and explain my findings to the jury, and the jury comes to a conclusion and renders its verdict.

In this book I am going to assume that you, the reader, are the Judge and the Jury. I must tell you what are my qualifications to be an expert witness on the subject matter of this book. You must judge whether or not I am qualified. I will then explain to you what are my findings, based on my engineering studies, and you will render the verdict. The subject of the case involved here is the origin of animals on this earth. I must prove to you that animals are machines, and that I am an expert in the design and origin of machines.

It is so important that you understand the overall format which organizes what is to be revealed in this book, that I feel it would be prudent for me to tell you of a couple other law cases on which I served as an expert witness. Here are the facts concerning another case.

Ammonia Tank Explodes

On May 17, 1973 a farmer was applying liquid anhydrous ammonia fertilizer to his land, near St. Henry, Ohio, at night. He had mounted on his tractor two large tanks of liquid anhydrous ammonia, one on each side of the engine. When he failed to return to the house for breakfast the next morning, his wife went out to the field and found him dead, lying on the ground. One ammonia tank had exploded. There were no witnesses to the accident!

Tractor with AmmoniaTanks

I was retained as an expert witness to investigate the accident. I found that there were two causes for the explosion: (1) the tank was overfilled with too high a percentage of liquid ammonia vs gaseous ammonia, and (2) the safety pressure-release valve on the tank was not properly maintained and it was stuck shut. Then, as heat radiated from the tractor engine, it caused the liquid ammonia to expand, absorbing the vapor phase, until the tank was completely filled with liquid. Further expansion, with no safety release, caused the tank to explode. The explosion knocked the farmer off of the tractor, and the liquid ammonia then vaporized, filling the surrounding region with deadly ammonia vapors. The combination of the explosion, knocking the farmer off of the tractor, and the toxic fumes caused the death of the farmer. The persons responsible for the accident were the man who filled the tanks, and the persons who were responsible for the maintenance of the relief valve. I gave my expert testimony. I explained what happened and who caused it to happen, based on my engineering studies. The farmer's heirs received appropriate compensation. You should now understand clearly what an expert witness is, and what he does. He is a person who, through education and experience, is a highly qualified expert in some field of knowledge. In a lawsuit, he studies matters related to his specialty. He reports his findings to lawyers before the trial, and then he testifies in court, after the judge agrees that he is a qualified expert. He tells the jury what happened and who caused it to happen. The jury then reaches a verdict.

Propane Truck Explodes

Propane Truck

Let's now consider two more cases. On March 13, 1979 a truck driver parked his truck in front of his mother's home in Troy, Alabama, and he went into the house to eat his lunch. The truck was a tanker loaded with 2000 gallons of liquefied petroleum gas. Shortly there-after the tank on the truck exploded, and the gas caught fire. The huge tank then became a rocket. It ripped itself off of the truck and was propelled at high velocity between two houses. Then, apparently, it began to spin in a back yard and it eventually came to rest there. A 17-year old boy, who was apparently standing near the truck, was blown into a nearby cemetery. He died. The blown-out front of the tank was found 239 feet away, and various parts of the truck were found hundreds of feet away from the parked position of the truck. The entire neighborhood was filled with propane gas, on fire, and the interiors of many houses became fiery infernos. But there were no witnesses to the actual explosion!

I was retained to investigate the accident and determine, as usual, what happened and who caused it to happen. I reported my findings to the legal people involved. Again, the tank was overfilled, this time with explosive liquid-petroleum gas, and the safety valve was improperly maintained. It was stuck shut. The original paint on the valve knob had never been disturbed. The persons responsible for the filling and the maintenance caused the explosion.

The Cambrian Explosion

The
Cambrian
Explosion

Of course, all of my law cases have not been related to explosions, although the above did all involve explosions, and there were no witnesses. Let's now consider a case in which there was an explosion, and it was a much greater explosion than any of the above. According to biological and earth scientists, about 540,000,000 years ago, there was a tremendous explosion on the earth. It is refer-red to by scientists as the "Cambrian Explosion." (Nash 1) It was an explosion in the sudden appearance of animals on the earth. Prior to the Cambrian period, the fossil records seem to indicate that there were few, if any, animals on the earth. Then, all of a sudden, the earth exploded with the appearance of a tremendous variety of animals, many of which have descendants living on the earth today. There were no witnesses to this explosion who could testify today.

So I, as an expert witness, have accepted the challenge of attempting to determine what happened, and who caused this explosion to occur. But first, of course, I must convince you, the Judge, that I am qualified to study this explosion and report on my findings. I must convince you that I am qualified on the basis of education and experience to be considered an expert in my field, which is the engineering design of machinery. And I must then convince you that animals are machines. I must convince you that I am a qualified and experienced expert witness in law cases. Then I must convince you, in your role as the Jury, that my findings are the truth. These matters will be discussed in the next Chapter.

CHAPTER 3. THE QUALIFICATIONS OF OUR EXPERT WITNESS

Walter L. Starkey

I must now tell you, the Judge, what are my qualifications to be an expert witness in my field. The following facts are offered here for qualification purposes only, and not for any purpose of self aggrandizement. An expert witness must reveal his qualifications as fully as possible to the judge, but he can still offer his qualifications with a genuine attitude of humility. I want that to be my attitude.

Education

First let's talk about education, then professional experience. I graduated from the Lyons Township High School in LaGrange, Illinois. My grades were near the top of the class of several hundred students. I then attended Lyons Township Junior College, in LaGrange, Illinois, for one year. I was a pre-engineering student. I was the only male student in the College who made straight A grades that year. I then attended, and graduated from, the College of Engineering of the University of Louisville. My major was mechanical engineering. My grades there, through three years of college were almost straight A's. In spite of having health problems and having to stay out of school for a year due to illness, my grade-point average at the University of Louisville was 3.86, where 4.00 would have been perfect.

I then attended The Ohio State University, and, in one year, I earned the Master's degree in mechanical engineering. My grades were straight A's, a 4.00 average. Finally, I attended Ohio State for several more years and earned the Doctor's degree, the PhD degree. I specialized in mechanical engineering, and more particularly, in machine design. My grades in the Doctoral program were all straight A's.

Professional Experience

My professional experience can be summarized as follows. Basically, I spent about 70 percent of my professional career as a college professor, and 30 percent as an expert witness, serving the legal profession. I was employed for four years by the University of Louisville, and for 32 years by The Ohio State University. As a college professor I developed and taught undergraduate and graduate courses. I also did research, published papers, and served as a consultant to industrial companies, government agencies, and to the legal profession. After retiring from being an active professor I devoted my full time to serving as an expert witness. I am still a professor. I hold the title, professor emeritus, at The Ohio State University.

Teaching and Research in Machine Design

My field of specialty is machine design, a subtopic of mechanical engineering. I taught students to design machinery. During my career as a professor I taught graduate and undergraduate courses in machine design, I directed 45 Master's-degree theses and 12 PhD-degree dissertations. As a research scientist, I performed research for government agencies and industrial corporations, including the Air Force, Army, National Science Foundation, Atomic Energy Commission, Public Health Service, Department of Transportation, General Electric Corporation, Ford Motor Company, Allis Chalmers, Owens Corning, Kelsey Hayes, Dayton Rubber Company, Boeing, Rockwell and many others.

> 24A **Columbus Dispatch** • • • • FRI., JUNE 17, 1966
>
> ## STUDIED FOR FORD
>
> # Brake Lining Test Simplified at OSU
>
> A quality control test for automobiles brake linings of automobiles, which duplicates in about 12 minutes the braking power used in thousands of miles of driving, has been developed by a team of Ohio State University mechanical engineers.
>
> THE WORK WAS performed at Ohio State by Prof. S. M. Marco, chairman of the department of mechanical engineering; Truman G. Foster, project supervisor; and Walter I. Starkey, professor.
> Their solution to the problem was an apparatus they have called the "ring-spot"

Publications

I published in major professional journals 27 publications, all on the subject of machine design, how to design machines so that they will perform and not fail. The following list identifies the subject matter of some of my publications: design to prevent failure by metal fatigue, machine failure theories, a new complex stress-time fatigue machine for aircraft propellers, a new design concept for machine members, a double-eccentric cam, high-performance machine design, statistical analysis of aircraft propeller steel, the invention of artificial heart valves, energy of recoil in army guns, fretting fatigue of Titanium in jet engines, stresses in wire ropes, design to prevent fatigue failure of V-belts, V-belts on locked-center drives, high-stress fatigue in jet engine compressor blades, the design of a cardiac simulator, friction instability in automobile brakes, corrosion fatigue, etc.

Consulting Practice

Throughout my professional career I have been a design consultant to many industrial companies, especially in the high-tech aerospace industry. Among the projects for which I have been a designer or a design consultant are the following: Bell Laboratory's radar antenna systems, the Atlas satellite booster, RCA's radar system for satellites, Aerojet General's satellite booster, the Lockheed Electra aircraft, Fairchild's VTOL aircraft, the Nike-Zeus missile system, Grumman's Roto-dome radar unit for Navy aircraft, Chrysler's 600 HP bridge-ferry assault ship, Lockheed's rigid-rotor helicopter, the Dynasoar reentry vehicle, Bell Aero System's VTOL aircraft, Lockheed's C5A Jumbo-jet aircraft, Lockheed's C-141 Transport aircraft, and many others.

Awards

Ohio State Unive

confers upon

Dr. Walter L. Starkey

the title of

Distinguished Alumnus

of his eminent contributions to profession and related field

In 1965 I received the Habel Award from Ohio State for spending two years chairing a committee and writing two books on the subject of the future of engineering education in America. In 1967 I was voted by the Ohio State faculty to be the chairman of the Faculty Advisory Committee to the President and The Board of Trustees of the University. Ohio State has about 5000 faculty members. In 1967 I served as National Chairman of the Machine Design Division of the American Society of Mechanical Engineers. This is the top administrative position in my Division of the Society.

In 1971 I was the recipient of the Machine Design Award, given by the Machine Design Division of the national level ASME. This is the top-level award in my profession. In fact, in my field, this is the equivalent of the Nobel Prize.

In 1996 I was given the very prestigious Distinguished Alumnus Award by The Ohio State University, for distinguished professional achievement in the field of engineering and machine design.

Expert Witness in Law Cases

During my career as an expert witness I have worked on about 150 law cases, many of them being multi-million-dollar cases, all on subjects related to the field of machine design. Each of the cases required engineering studies of the facts involved in a lawsuit. Most of the cases involved accidents, or the failure of machine components. But some of them involved alleged theft of trade secrets or patent disputes.

Some of the machines involved in these cases include the following: motorhomes, autos, trucks, tractors, helicopters, aircraft, cranes, aerial-lifts, mining-machines, punch-presses, wood-chippers, propane tanks, "steam" rollers, air compressors, ammonia tanks, motorcycles, road-graders, bull-dozers, conveyors, motor-generators, etc. Engineering studies were made in all cases. For most of the lawsuits, prior to the trial, I was deposed by lawyers. They asked the questions, and I told them what happened, and who caused it to happen. It was my role to determine who was responsible for what happened. Many of the cases went to trial in front of a judge and jury.

Was I a successful expert witness? Yes I was! It is a fact that my side was the winner in almost all of the cases in which I participated. In one case, the judge told the lawyer with whom I was working that I was the best expert witness he had ever seen.

Scientist, Engineer, and Professor

One topic having to do with qualifications is an interesting one. Who is to be held in the highest esteem, a scientist, an engineer, or a college professor? In my case, it doesn't matter because I am all three. A scientist is one who endeavors to discover new knowledge. An engineer is one who applies scientific information and designs products that are useful. When I am doing research, I am a scientist. When I am consulting with industrial companies, or designing machinery, I am an engineer. And, when I am teaching students how to design machinery, I am an educator, a college professor.

Qualified as an Expert Witness

Again, may I emphasize that all of the above information is offered to you, the presiding Judge, with the hope that you will be convinced that I am qualified to be an expert witness in your court room. More than 150 judges have already concluded that I am qualified. After learning the facts, no judge has ever questioned my qualifications. But, most of all, I want you to believe that what I say is the truth. All of the above are facts, and they are stated only for the legal purpose of being qualified, certainly not in any spirit of arrogance. It is important for you, the Reader and the Judge, to become better acquainted with me, the Author and the Expert Witness.

CHAPTER 4. ASSIGNMENT: WHO CAUSED THE CAMBRIAN EXPLOSION?

In Chapter 1 of this book we explained that we would use the lawsuit format within which to develop the material of the book. In Chapter 2, we explained how an expert witness operates. He performs investigations, arrives at conclusions, and testifies concerning his findings in court. In Chapter 3, I presented my qualifications to be an expert witness in the field of machine design. In this Chapter we will assume that I am qualified, and that I have been given the assignment to study the Cambrian Explosion, and determine who caused this Explosion. This, of course, is the equivalent of determining who designed and constructed the animals of the earth. We will now explain in more detail what is meant by the Cambrian Explosion, we will attempt to establish when it happened, and we will begin to show what might be involved in determining who caused this explosion.

History of the Cambrian Explosion

According to scientists, the universe began with the sudden appearance of matter and energy about 15 billion years ago. Also, according to scientists, the earth was formed about 4.6 billion years ago. Scientists have defined a series of time periods which cover the entire history of the earth. These are used to identify the periods during which various geological and biological events took place. One of these time periods is called the Cambrian period. It began about 0.6 billion (600,000,000) years ago and it lasted about 0.1 billion (100,000,000) years. Based on the findings of paleontologists and geologists, the earth was essentially devoid of animals for about seven-eighths (7/8) of its existence, from its origin 4.6 billion years ago to 0.6 billion years ago. Then, within the Cambrian period, thousands of species of animals suddenly appeared. Recently some scientists have come to the conclusion that this explosion occurred about 540,000,000 years ago, and that it lasted only about 5,000,000 years. This sudden appearance of animals is called the "Cambrian Explosion."

Books on biology describe this Explosion. Typical is the following account, written by Hickman, Roberts, and Hickman, in their book, *Biology of Animals*. (Hickman 1):

> "At the beginning of the Cambrian period, most of the major phyla of invertebrate animals made their appearance within a few million years. This has been called the 'Cambrian explosion' . . . The burst of evolutionary activity that followed at the end of the Precambrian period and beginning of the Cambrian period was unprecedented; nothing approaching it has occurred since. Nearly all animal and plant phyla appeared and established themselves within a relatively brief period of a few million years."

Now, the question to be answered by this book is, "Who caused the Cambrian explosion?". There were no witnesses to it who are alive today, but that causes no problem for a properly qualified expert witness. Many of the explosions and other accidents which I have investigated as an engineering expert had no witnesses. The challenge of determining who did it, in the case of the Cambrian explosion, is certainly not unique in my experience.

Theories Concerning Who Caused the Cambrian Explosion

So who did cause the Cambrian explosion? Who created the animals? As we stated in Chapter 1, there are basically two general theories which attempt to answer this question: (1) the theory of evolution, and (2) the theory of creation. There are several specific theories of evolution, but all of them can be analyzed as consisting of two phases: (a) Some natural forces, by chance, act on an individual animal, which cause it to have some new feature. This feature may either enhance or diminish the ability of that animal to survive. Then the second phase occurs. (b) The animal will be acted upon by natural-selection. An animal that has an enhanced ability to survive will tend to continue living and it may produce offspring having the same advantageous feature. All others will have lesser ability to survive, and they will tend to become extinct. Thus, the general theory of evolution consists of some New-Feature-Producing Agent, which might produce a new feature, and which we will call an NFPA, followed by the second phase, Natural Selection.

Specific theories of evolution differ one from another in the identification of a particular new-feature-producing agent. Candidates for this role include such entities as Mother Nature, Teleology, Adaptations, Mutations, Mendelian-genetics, and Punctuated Equilibrium. In addition, all of these theories involve the second phase, natural-selection, acting on the new features. Adherents of all of the theories of evolution believe that the animals came into existence by the actions of the natural physical, chemical, and nuclear forces of the universe, alone, without any assistance from any supernatural being. This means that evolutionists believe that the animals were designed and constructed by nobody.

It is clear that the process of natural-selection does definitely take place in nature. Since the Cambrian Explosion, thousands of species have become extinct. We know, today, of many species which are on the verge of becoming extinct. The concepts of natural selection and the survival of the fittest are readily believable. But the first phenomenon, the creation of new animals by the chance actions of natural forces, as we shall see later in this book, is much less believable.

According to the theory of creation, all of the animals of the earth were created by some extremely competent supernatural designer and builder.

How to Carry Out Our Assignment

In order to carry out this assignment, we intend to employ the same procedures that we would use in any typical law case involving machinery. We will study both the theory of evolution and the theory of creation. First, we will objectively and thoroughly study the theory of evolution, and, if it should prove to be the truth, we will accept it, and conclude that the animals were, indeed, designed and constructed by no-body, just the natural forces of the earth. However, if we should conclude that the theory of evolution is not the truth, and that the animals were created by somebody, we will attempt to determine just who that somebody might have been. If that somebody has an identity and a name, we will attempt to find out who it was. This is not a religious book. Rather it is a book which employs scientific facts, engineering analyses, and is based on historical records.

Summary

In summary, scientists tell us that, about 540,000,000 years ago, all of a sudden, animals came into existence. This is called the Cambrian Explosion. We want to find out who caused this Explosion, who created these animals. Some believe that a supernatural being designed them and built them. Others believe that the design and construction of animals came about by impersonal natural forces. The expert-witness assignment that we have accepted is to determine if the animals of the earth were created by nobody, or by somebody; and if we determine that they were created by somebody, we want to find out who that somebody was.

CHAPTER 5. FINGERPRINTS OF A DESIGNER

Fingerprints Find Killer of Young Woman

Fingerprint

Footprints and fingerprints often provide important clues which can lead to facts and opinions which, in turn, can produce supportable conclusions relative to who is responsible for certain actions. Thirty-five years ago a young man and his very attractive young wife built a house in the country north of Columbus, Ohio. They enjoyed country living. The young wife was pregnant. One evening the young man came home and found his wife dead; she had been brutally murdered. There was snow on the ground outside, and footprints lead detectives to the home of a suspect. Fingerprints inside the victim's house convinced the detectives that the suspect did, indeed, commit the crime. He is now in the Ohio penitentiary.

Fingerprints will be used to Find Who Caused the Cambrian Explosion

The purpose of this book is to perform the detective work necessary to determine who designed and constructed the animals on the earth. Were they designed by somebody, or by nobody? If they were designed and built by nobody, then they had to be produced by the natural forces of the earth, such as: wind, rain, hail, flooding, freezing, thawing, earthquake, lightning, ice movement, cosmic rays, etc. In Chapter 9 we will determine the statistical probability that a fire-ring of rocks could have been made by nobody. This probability will turn out to be one chance in 8.32×10^{33}. In Chapter 10 we will show that the probability that a protozoan could have been made by nobody would be one chance in 5.52×10^{49}. These determinations will be made by me, an engineer. I will conduct investigations as an expert witness, based upon my expertise in the field of engineering and machine design; and I will testify to you, the Judge and Jury, that there was essentially no chance whatever that the fire-ring or a protozoan could have been designed and constructed by nobody. In later Chapters of this book we will consider other animals of various types and sizes, and I will testify concerning their origins. The purpose of this Chapter is to inform you more fully concerning how I arrive at my conclusions. Upon what bases do I make my judgments, as an engineering expert? What are the footprints and fingerprints that I look for to determine the truth? Hopefully, these added insights will increase your confidence in my conclusions.

I will now identify the ten most important fingerprints, clues, which I look for when analyzing a machine, including animal-machines, to determine the characteristics of the machine, including its probable origin. These ten telltale fingerprints are explained below.

Ordered Arrays of Materials

The natural forces of the earth tend to produce a random and disorganized placement of materials. Figure 5.1 shows an ordered array of rocks. These were found in the Bighorn Mountains west of Sheriden, Wyoming. This group of rocks is called the Medicine Wheel. It consists of an array of rocks formed in the shape of a wheel, having a rim and 28 spokes. Nobody knows how it got there. The Crow Indians were the first people known to have occupied that area, and they say it was there when their ancestors arrived in the region. Some of the Indians, to this day, believe that it was designed and constructed by the Great Spirit; but most of them believe it was built by the human beings who occupied the area many years before their ancestors arrived. None of the Indians believe it came into existence by evolution, involving only the natural forces of the earth. What do you think?

Medicine Wheel Array of Rocks   **Figure 5.1**   Actual Rocks of the Array on Medicine Mountain in Wyoming

What other ordered arrays of materials might we examine? The wheels on my Jeep are attached to the hub by an ordered array consisting of five bolts in a circle. Also, on the Jeep wheel-hub is the brake disk with its fan-blades arranged with regular periodicity. The teeth of gears exemplify an ordered array. The wires of a wire rope are an excellent example of an ordered array. These machine parts were designed by human beings. They could not have come into existence just by the natural forces of the earth.

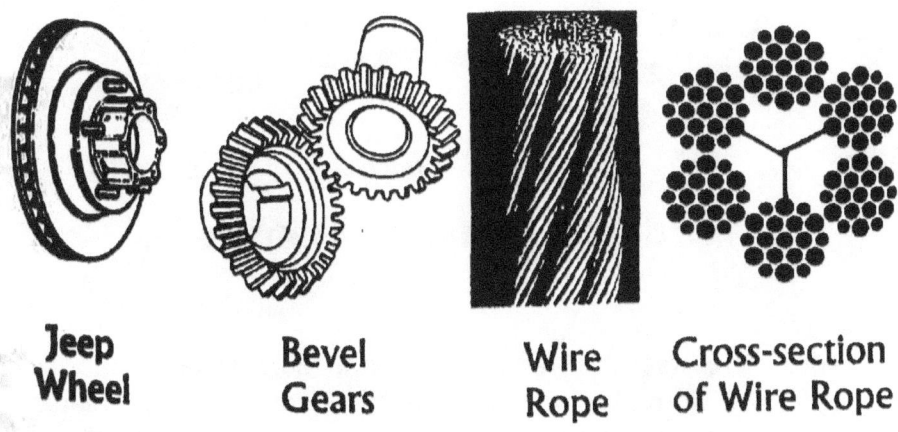

Jeep Wheel   **Bevel Gears**   **Wire Rope**   **Cross-section of Wire Rope**

Ordered Arrays in Machinery

The keys of a typewriter are arranged in an ordered array. Figure 9.1 of Chapter 9 shows a ring of rocks on a mountain top, which we will study. These are an ordered array. In Chapter 10 we will study an ordered array of nine pairs of microtubules in the flagellum of a protozoan. These are arranged in a circle, as shown in Figure 10.2. The bones of our spine constitute an ordered array. Our teeth are arranged in an ordered array. The following, also, are ordered arrays: the eight eyes of a spider, the bilateral symmetry of most animals, the fins of a fish, the legs of a centipede, the arms of a starfish, the four wings of a dragonfly, the pattern of colors on a butterfly's wings, the segments of an earthworm, the compound eyes of insects, the scales of a fish, the pattern of a turtle's back, the barbs of a bird's feather. These are all ordered arrays.

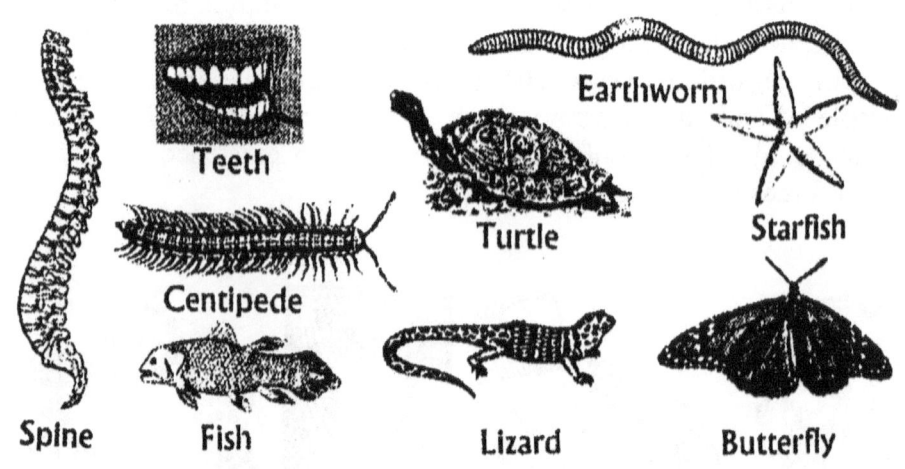

Ordered Arrays in Animals

Before there was any life on the earth, the surface of the earth consisted only of rocks, air, and water. There were no organized patterns; no ordered arrays of materials. When I see an ordered array, I conclude that it is the result of the design and construction efforts of some intelligent being. An ordered array is a fingerprint of a designer.

Shapes of Parts

Each of the parts of a machine, or an animal, has a shape. The following shapes strongly indicate that the part was designed and constructed by some intelligent being; smooth surfaces, straight lines, squares, rectangles, circles, triangles, planes, prismatic bars, sheets, hexagons, cones, spheres, hollow tubes, wires, spirals, helices, involutes, etc. The natural forces of the earth don't form rocks into these shapes. The natural forces of the earth don't refine metals from ores, and fabricate them into hollow tubes, round bars, thin sheets, helices, cones or any of the others of these shapes.

Consider a bolt and a nut. The head of the bolt is a hexagon made up of planes, and the beveled edges are cones. The shank of the bolt is a cylinder. The threads are helical with a triangular cross-section. The nut is made up of planes, a hexagon, cones, a helix, and triangles. Furthermore, the bolt and nut have a purpose. I have never seen a rock shaped like a bolt and nut. I say that the bolt and nut were designed and constructed by some intelligent being, a human being.

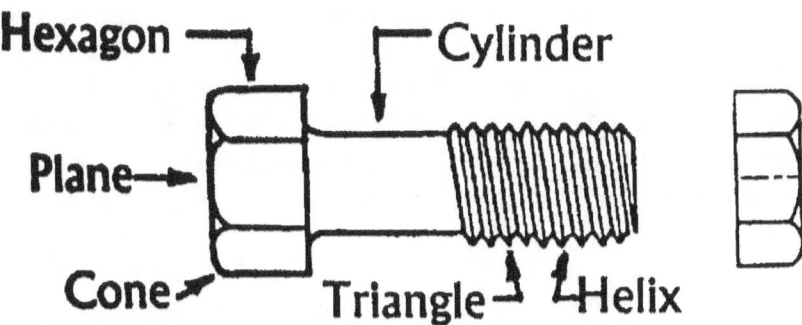

Nut and Bolt Show Shapes of Man-made Parts

When I study an animal, and note that it is filled with hollow tubes, for transporting blood, and it has wires, for carrying nerve impulses, and spherical ball joints in its hips, and spherical eyes that can rotate about any axis, and long straight bars in its legs, I conclude that it was designed and constructed by some being. These shapes are not produced by wind, rain, hail, lightning, earthquake, cosmic rays, or any of the other natural forces of the earth. Such shapes are produced by the machine tools of men, or by the skilled hands of a supernatural designer-craftsman.

Certainly, the shapes of parts are important fingerprints which suggest a designer.

Refined Materials

With the exception of gold and a few other precious metals, all of the materials used in man-made machines are found in the earth's crust in the form of ores, which are chemical compounds which contain desired metals; or oil, from which plastics can be made. These minerals must be mined and then refined by heat and/or chemical reactions before the useful materials can be obtained. Then additional chemical processes are often needed to produce alloys such as steel, stainless-steel, bronze, or specific plastics. In the case of animals, they eat food, and this must then be refined by their bodies, through extremely complex chemical processing, to produce the materials for their bones, teeth, organs, skin, muscles, nerves, etc.

When I see the refined materials in a machine, I know what processes have been required to produce these materials, and I know that these materials have been mined, refined, alloyed and processed by intelligent human beings. And when I see the materials of which an animal is made, and when I study the extremely complex chemical reactions and processes that are needed to produce these materials from the food the animal eats, I conclude that none of the natural forces of the earth could produce these materials. They had to be produced by complex processing devices designed and constructed by some intelligent supernatural being.

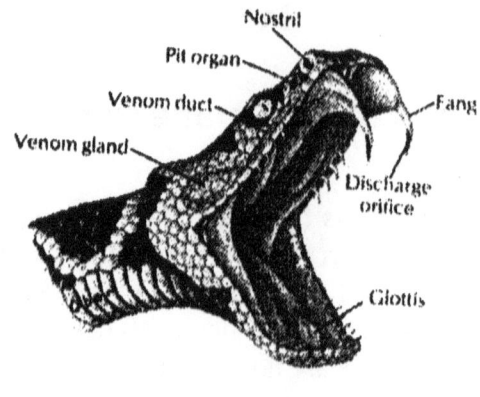

Nostril
Pit organ
Venom duct
Venom gland
Fang
Discharge orifice
Glottis

Do you think that, merely by chance, without a designer, a snake could eat a mouse, and turn this food into a deadly poison in a poison sack attached to a hypodermic needle? (Hick-man 13) Do you think that, merely by chance, a brown cow could eat green grass and turn it into white milk, one of the best foods on earth? Do you think that, merely by chance, a boy could eat a soft hamburger with some milk, and turn them into the enamel of his teeth, one of the hardest substances found in nature? Refined materials in animals clearly constitute another one of the fingerprints of a supernatural designer.

Manufacturing Process

Men Make Machines

Much can be said about the manufacturing processes human beings use to produce machinery, but little is known about the processes by which animals were made. Man-made machine parts can be manufactured by sand-casting, die-casting, forging, or by the machining of stock shapes. As an engineer, I can analyze machinery on the basis of the manufacturing processes used, and I can look into the mind of the designer by studying these processes.

Obviously, animals are not made by the use of machine tools. It would appear that, in general, the original designer-craftsman of each animal made a pair of them, male and female, and they were made in such a way that, through sexual intercourse, offspring could be created which, in turn, could create more offspring, and this process could continue indefinitely until the species became extinct. When the sperm of the male meets the egg of the female a new animal is created. The fertilized egg of the new animal contains DNA molecules which then direct the development of the new animal in every detail. The DNA serves as the plans, the detailed drawings, and the supervision needed to create the new animal. So, the manufacturing processes for animals, after the creation of the first pair, consist of cell multiplication guided by the DNA of the animal. The DNA supervises the manufacture of the animal, but who made the DNA?

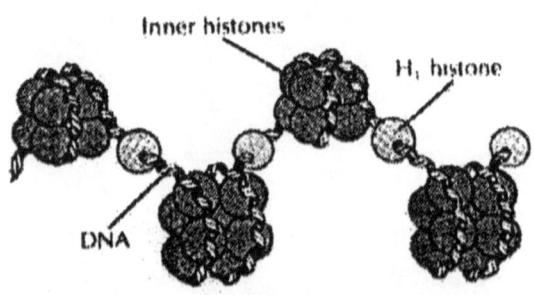

Inner histones

H₁ histone

DNA

DNA Supervises the Manufacture of Animals

The question here is, what chance is there that this manufacturing process could have come into existence by itself, powered only by the natural forces of the earth. My expert-witness testimony on this subject is that there is no chance that an automobile, aircraft, or other man-made machine could come into existence by the actions of such forces, and the manufacture of any animal is almost infinitely more complicated than the manufacture of any man-made machine. Therefore, it is inconceivable that any animal could be manufactured except by the deliberate efforts of an intelligent supernatural being who is not only skilled in design, but is also skilled in construction and manufacturing.

Multipart Systems

Let's consider a couple of specific examples of multipart systems. We will describe them briefly here. A multipart system is a group of separate parts of a machine or an animal which work together to perform a particular important function. A multipart system is like a steel chain having many links, used to pull a load. If one link is missing, the chain is worthless. If any one part of a multipart system in an animal is missing, the system is worthless.

First, let's consider the sex organs of mammals, including humans. This system has many parts. The chain of essential links in this multipart system even extends beyond the confines of a single individual. It involves two individuals, a male and a female. In the male, the parts include two testicles, seminiferous tubules within each testicle, tubes called vasa-efferentia, tubes called ductus-epididymis, tubes called vas-deferens, storage sacks called seminal-vesicles, a prostate gland and two Cowper's glands which secrete essential fluids, an ejaculatory duct, a duct called the urethra, and a penis, complete with its erectile system of nerves, muscles, and blood vessels. In the female, continuing with the same system, the parts include a vagina, a clitoris, glands which secrete lubrication, two ovaries, fallopian tubes, a uterus in which a fertilized egg can develop into a new individual, and associated nerves, blood vessels, and birthing muscles. If any one or pair of the above-identified thirty parts were missing, the system would not work. If the penis were missing, it wouldn't work. If the vagina were missing it wouldn't work. It needs the testicles and ovaries. The system is filled with tubes. Even if evolution could produce one complex part of a multipart system, what statistical probability is there that it could simultaneously produce thirty complicated parts, all of which are required for the system to operate at all. When I see just one tube, I see a designer. Rocks don't produce tubes. When I see thirty complex parts that must work together, I know there was a designer.

Do you think that a single chance mutation, or even a series of them, could create a complex system like this? Do you have some other reasonable explanation as to how this system came into existence? I can visualize a chance mutation causing a slight bump on a man's leg, but could that serve as a penis plus 30 other complex parts? To me, such thinking is ludicrous.

Consider next the multipart structures and mechanisms which enable a bird to fly. A typical modern small airplane has two wings, several rotating propellers, a fuselage, and tail surfaces which provide flight control. The inner portions of a bird's wings move only slightly and serve as aerodynamic supporting surfaces. The outer portions of a bird's wings serve as reciprocating propellers. Instead of rotating, they move up and down, but they also rotate on their axes during each cycle so there is for-ward thrust whether the wing is going up or going down. The body of the bird is its fuselage, and its tail feathers provide for flight control.

Just as for airplanes, birds must be light in weight, and have powerful engines of propulsion. The bones of birds are not only hollow, but they are also reinforced within by short bars to prevent elastic-instability-type failure of the thin walls of the hollow bones, a remarkable engineering structure! The frigate bird has a wingspan of seven feet, but the total weight of all of its bones amounts to only four ounces. Birds have a sternum bone made in the form of a very deep keel to which are attached the large and powerful muscles which move the wings. The large lower muscles pull the wings down. Above these are the slightly smaller muscles which, through

Airplane has Multipart Systems

Bird has Multipart Systems

a pulley-tendon system, pull the wings up. The airfoils of the wings and tail are made up of feathers. Only birds have feathers. These are highly efficient flat structures consisting of a central quill and shaft, with barbs and barbules extending outward on either side of the shaft. See Figure 12.2. The barbules lock the barbs together to form a strong and rigid airfoil. One feather may have as many as a million barbules. Birds even have spoilers to reduce the speed at which they would engage in aerodynamic stall. A bird is made up of many parts. But it wouldn't be able to fly if any of the parts mentioned above were missing. If a bird didn't have hollow bones, it would be too heavy to fly. If it didn't have a tail, it couldn't fly. The Wright Brothers found that an airplane needs a tail. If a bird didn't have its deep keel and strong breast muscles, it couldn't fly. Birds have many multipart systems, all designed to enable them to live and fly.

Evolutionist believe that birds evolved from reptiles, by mutation and natural selection. I can envision a slight change occurring in a reptile due to a mutation; but I can't, in my wildest of dreams, imagine a bird, capable of flying, being produced from a reptile by chance mutations. The Wright Brothers first designed bicycles, and then airplanes. Do you think that the natural forces of the earth could turn a bicycle into an airplane without any help from the Wright Brothers? I think it would be equally ridiculous to believe that the natural forces of the earth could turn a reptile into a bird! If the smartest man in the world told you that birds evolved from reptiles, would you believe it?

There are literally thousands of similarly-complex multipart systems to be found in animals. The following list identifies a few typical examples of such systems: food-processing and digestive systems, optical systems

32

for sight, auditory systems for hearing, computer-like brains, nervous systems which transmit messages by electricity, muscular-skeletal systems which enable animals to move and exert forces, systems for poisoning prey, systems for getting oxygen out of sea-water, sonar systems to aid navigation, etc.

Mutations might produce slight modifications, but random, chance mutations cannot design and construct complex purposeful multipart systems. When I see such multipart systems in an animal I see an intelligent and knowledgeable supernatural designer behind the system. Multipart systems are among the most important fingerprints of a designer.

Complex Mechanical Systems

There are two types of mechanical systems: (1) solid systems, and (2) fluid systems. Many machines contain both solid and fluid systems. The power of the engine in a typical automobile goes through pistons, piston-rods, the crankshaft, the transmission, the drive shaft, the differential gears, and the axles to the driving wheels. These are all solid parts. The automatic transmission, the hydraulic brakes, and the air-conditioning system of an automobile involve fluid systems. A backhoe involves hydraulic cylinders driving rigid and solid parts. The muscle of an animal exerts force by shortening and pulling on a tendon. These are mechanical systems involving solid parts.

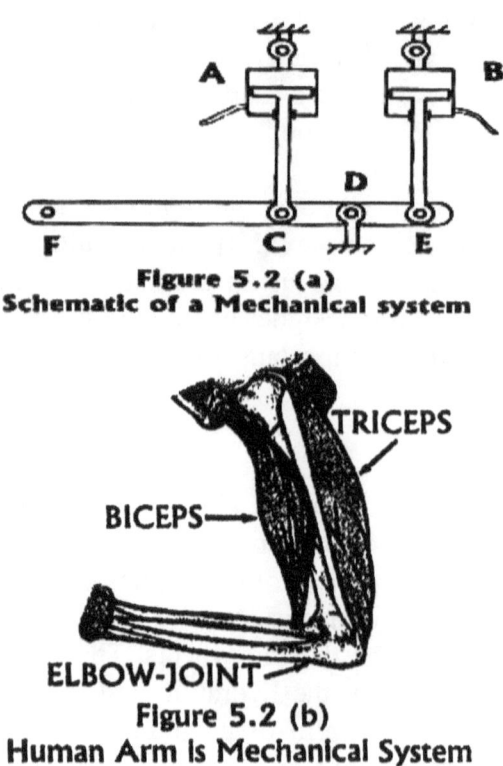

Figure 5.2 (a)
Schematic of a Mechanical system

Figure 5.2 (b)
Human Arm Is Mechanical System

Figure 5.2 shows a mechanical system. If hydraulic cylinder B, of Figure 5.2 (a), lifts joint E upward, the lever FE will rotate about pivot D, so that point F will go downward. If cylinder A be activated and it pulls joint C upward, then arm FE rotates in the opposite direction, and point F will move upward. Figure 5.2 (b) shows the muscles and bones of the human arm. They work in an identical manner. The triceps are analogous to cylinder B, and the biceps are comparable to cylinder A. The elbow joint is analogous to pivot D. The human arm is a mechanical system.

When I analyze a backhoe, or other similar machine, I can visualize the thought processes of the designer of the mechanism. I know that an intelligent skilled engineer designed the device. Similarly, when I analyze the mechanical system that is the arm of a human being, or the limb of an animal, I see irrefutable evidence of the existence of its designer.

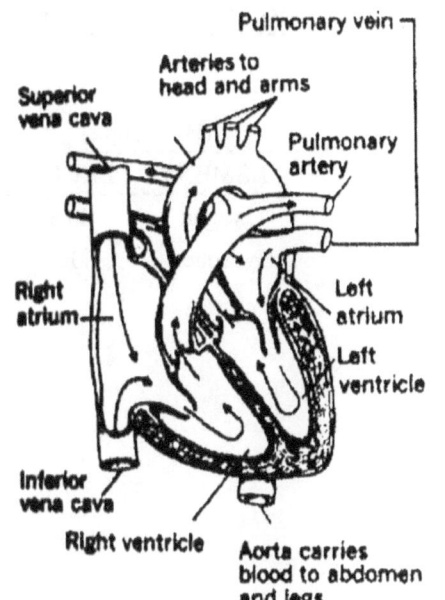

Animal Heart

Similarly, the four-valve heart and blood circulatory system of an animal is a mechanical system involving the engineering design principles of fluid mechanics. (Wagman 1) I have a water-treatment system in my boat that has four valves and two pumps. I know it didn't evolve. Somebody designed it. I also know that the four-valve two-pump heart in an animal is a complex mechanical system, and it was not designed by nobody.

Mechanical systems are certainly among the important fingerprints of a designer.

Complex Chemical Systems

Before we had plants and animals on our planet, all that existed on earth were rocks, water, and air. These, of course, can be thought of as chemicals. So we need to study a little basic chemistry. All substances of the earth are composed of atoms. Atoms can be combined to form molecules. This is accomplished by chemical reactions. To make animals out of the chemicals of the earth it is obviously necessary to apply chemical reactions to the chemicals of the earth so that they become recombined into the chemicals of the animals. So we need to study the chemicals of the earth and the chemicals in the bodies of animals.

There are 92 naturally-occurring elements, or kinds of atoms, on the earth, but only a few of them are present in large amounts near the surface of the earth. These are shown in Table 5.1.

TABLE 5.1. CHEMICAL ELEMENTS NEAR THE SURFACE OF THE EARTH

IN THE EARTH'S CRUST (By Weight)		IN THE ATMOSPHERE (By Volume)		IN THE SEAS (By Weight)	
ELEMENT	PERCENT	ELEMENT	PERCENT	ELEMENT	PERCENT
Oxygen	46.6	Nitrogen	78	Oxygen	88.9
Silicon	27.7	Oxygen	21	Hydrogen	11.1
Aluminum	8.1	Argon	0.93		
Iron	5.0	Carbon dioxide	0.03		
Calcium	3.6	Water Vapor	Various		
Sodium	2.8	Others			
Potassium	2.6				
Magnesium	2.1				
Others	1.5				

Somehow, some of these elements have become combined by chemical reactions, to form animals, including human beings. The elements in a human being, as well as many animals, are as shown in Table 5.2.

TABLE 5.2. CHEMICAL ELEMENTS IN A HUMAN BEING

ELEMENT	PERCENTAGE (by Weight)
Oxygen	85.00
Hydrogen	10.70
Carbon	3.30
Potassium	0.34
Nitrogen	0.16
Phosphorus	0.05
Calcium	0.02
Magnesium	0.01
Iron	0.008
Sodium	0.001
Zinc	0.0002
Copper	0.0001

Obviously, the animals are not just mixtures of the earth's crust, water and air, because the major constituents don't match. In the earth's crust there is a lot of silicon and quite a bit of aluminum, but animals have almost none of these elements. On the other hand, in the human body there is 3.30 percent carbon, but the percentage of carbon on earth is much less. In fact, it is about 0.000,005,53 %. This is close to zero. Let's calculate it.

About 99.9% of the earth's water is in its oceans. The oceans cover 71% of the earth's surface to an average depth of 2.3 miles. Water is 1000 times as heavy as air, and rocks are 2.6 times as heavy as water. The carbon-dioxide in the air constitutes only 0.03% of the air, and the carbon in carbon-dioxide is only 27%. So the carbon in the air is only about (0.03)(0.27), or 0.008% of the air. This is less than one part in ten-thousand. Based on these figures, the percentage of carbon near the earth's surface, measured for two miles up and two miles down, can be calculated to be about,

$$C\% = (0.03)(0.27)/ [(0.71)(1000) + (0.29)(2.6)(1000)],$$

$$C\% = 0.000,005,53\%.$$

This is the equivalent of one part in 18 million! But, in the bodies of animals there is 3.30% carbon. This means that, on a percentage basis, there is 597,000 times as much carbon in the bodies of animals, as on the earth. Thus, it would appear to be quite remarkable that so much of this scarce element, carbon, has found its way into the bodies of animals. Could it be that whoever designed the animals knew that the element, carbon, would be uniquely suited to the construction of animals?

Why is it that carbon is so important? One reason is that carbon is unusual because it is a lightweight element, but it has a high chemical bonding valence. Valence is a measure of the extent to which an atom can combine with other atoms. Hydrogen has a valence of 1. Oxygen has a valence of 2. But carbon has a valence of 4. Thus carbon has a strong tendency to combine with other elements, or with other carbon atoms. Carbon atoms, and other elements, can combine to form molecules which contain long chains of atoms, or rings of atoms, or sheets or large chunks. The resulting molecules can be very large and very complex. Many carbon-based molecules contain thousands of atoms, and some can even contain millions of atoms. Figure 5.3 shows the group of atoms which is the sugar, glucose. Figure 5.4 (Hickman 14) shows a combination of atoms, featuring carbon, which is lecithin. In animal's bodies, lecithin is an important ingredient in nerve membranes.

Lecithin is shown here to begin to demonstrate the great complexity that can be achieved in molecules based on carbon.

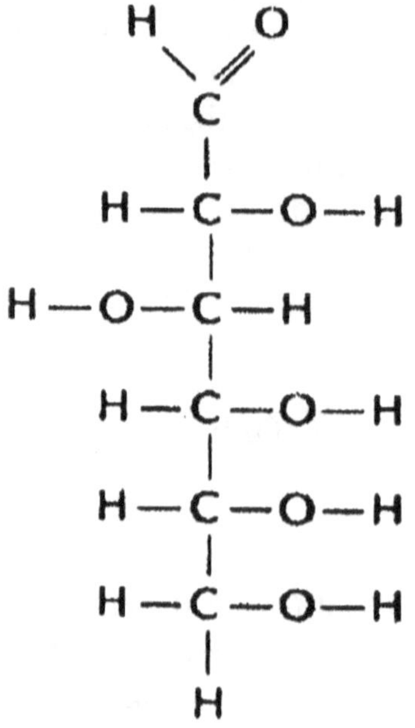

Figure 5.5 shows a combination of carbon, hydrogen, nitrogen, oxygen, and phosphorus which contains a sugar, a base containing nitrogen, and a phosphate group. This combination is called a nucleotide. Nucleotides can be combined to form very large molecules. One such molecule, called DNA, may contain as many as three billion such nucleotide groups. We will discuss this molecule in more detail later.

Glucose

Figure 5.3

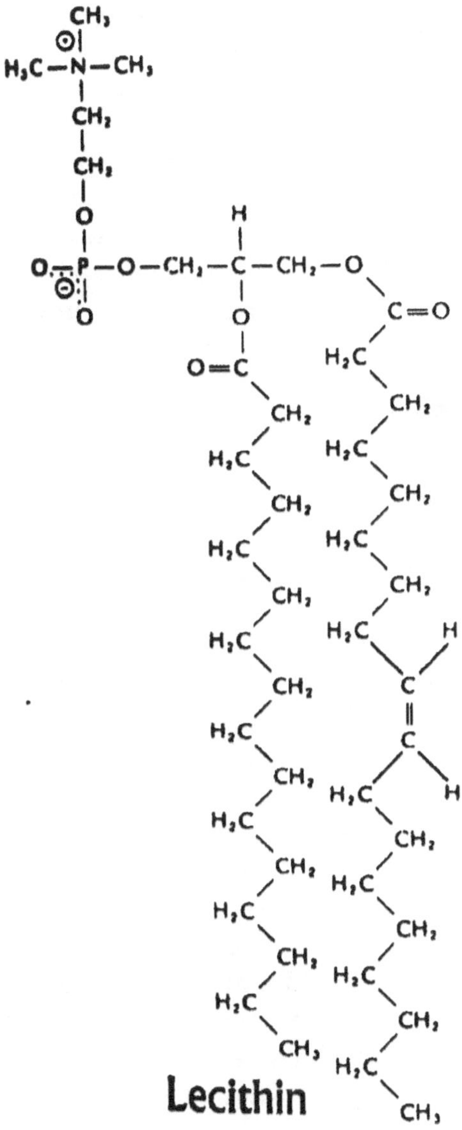

Lecithin

Figure 5.4

Obviously, we cannot delve deeply here into the mind-boggling chemistry of carbon-based molecules and their related chemical reactions, but, just to demonstrate the breadth and complexity of the chemicals in animals, we will simply name some of the carbon-based substances that are important to animals. The following are found in animal cells: sugars, fatty acids, amino acids, nucleotides, groups called methyl, hydroxyl, aldehyde, ketone, carboxl, and phosphates. Then we have carbohydrates, m o n o s a c c h a r i d e s , polysaccharides, lipids, glycerides, phospholipids, waxes, steroids, proteins, cysteines, glutamates, glycines, phenylalanines, trytophans, trosines, valines, adenosines, nucleotide-coenzymes, and many others. Some of the components found in living cells of animals include: ribonucleic acid, ribosome, endoplasmic reticulum, golgi bodies, lyosomes, mitochondrions, cyloskeletons, erthrocytes, leukocytes, thrombocytes, and others.

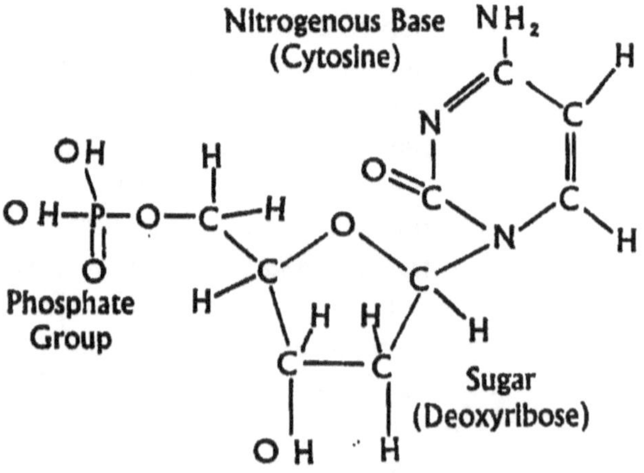

Nitrogenous Base NH₂
(Cytosine)

Phosphate Group

Sugar
(Deoxyribose)

Example of a Nucleotide as Found in DNA
Figure 5.5

Carbon-based compounds in living animals perform the following functions: they build structures, provide energy, store food, form cell membranes, provide enzymes, provide hormones, produce hair and nails, build bones, transport oxygen, translate genetic information, and perform many other tasks.

The above identified chemical compounds, biological entities, and functions, all based on carbon and organic chemistry, should convince you of the very great complexity of the chemistry of animals. Animals are certainly not just a mixture of the chemicals found near the surface of the earth. But they are, indeed, made up of these chemicals, carefully selected.

Obviously, the conversion from rocks, water, and air to animals must have involved many chemical processes. Somehow, the atoms in the rocks, water, and air were recombined into the atoms of the animals. The question is, how did this come about? It must have been done by one of the following three agencies: (1) by some supernatural person, (2) by the smartest creatures on earth, human beings, or (3) by nobody, involving only the random chance actions of the natural forces of the earth. Which of these three agencies really did do it might be suggested by a study of how hard it must have been to accomplish these conversions. How complicated are the chemical reactions that had to occur to effect such conversions?

Of the three agents, the least intelligent one is nobody, meaning evolution. Human beings are much more intelligent than nobody. And our supernatural being must be, by far, the smartest of the three. You, dear Reader, are a human being. Let's see if you are smart enough to bring about these chemical conversions. If you can't do it, then, surely, evolution couldn't do it, and we will have to conclude that it was done by the supernatural being. So, let's consider what you could do. The first step in creating any animal would be to bring about the chemical reactions by which some of the rocks, water, and air were chemically converted into DNA molecules for the selected animal. Every animal needs DNA molecules to provide the plans and the supervision needed for its development. This includes all animals, from a protozoan to an elephant. So let's now determine whether or not, based on your knowledge of chemistry, you could create just one molecule of DNA.

A DNA molecule contains carbon, oxygen, hydrogen, nitrogen, and phosphorus. These must be chemically combined to form nucleotides, such as shown in Figure 5.5. Then about six billion of these nucleotides must be chemically combined in the form of a double helix, in a particular order and in a particular orientation. In each nucleotide there are about 34 atoms, so you will need to chemically combine, from the rocks, water, and air, about (34)(6,000,000,000), or 204 billion atoms. If you could process atoms at the rate of one per second, and if you worked 8 hours per day 350 days per year, it would take you more than 20,000 years just to produce one molecule of DNA. I conclude that you couldn't produce even one such molecule, and hence you couldn't create any animal, not even one protozoan. If you, a smart human being, couldn't do it, what statistical probability do you think there is that one molecule of DNA could be produced by nobody, by evolution? I am of the opinion that there is no chance! And, without any DNA molecules, no animal of any sort could be created or exist!

As an engineer, and as an expert witness, when I see complex chemical compounds, chemical reactions, and chemical systems, such as those described above, in the bodies of animals, I see the hand of a very intelligent, very knowledgeable, very skilled designer and craftsman, and a very skilled chemist. Chemical systems are among the important fingerprints of a designer that can be found through a study of animals.

Complex Electrical Systems

Another type of complex system that I look for as I analyze machines, including animal-machines, includes those which are based on electrical phenomena. All animals have nerves and neurons, which are electrical devices, and the neurons of all animals are of the same design. So, let's study neurons.

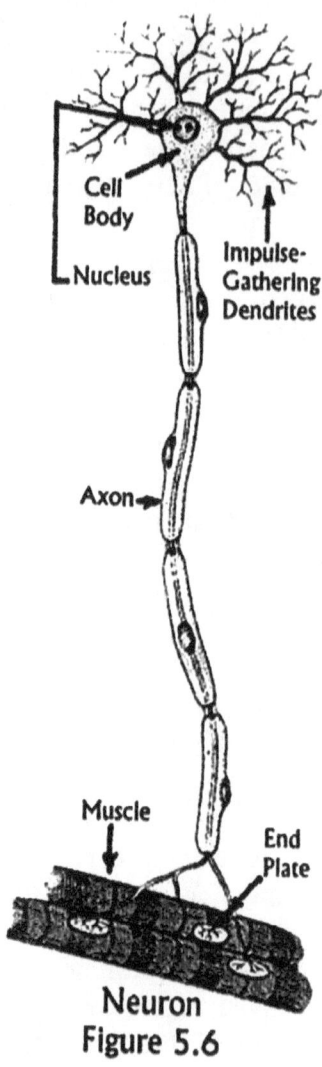

Cell Body

Nucleus

Impulse-Gathering Dendrites

Axon

Muscle

End Plate

Neuron
Figure 5.6

Man-made machines are built largely of metals, and their electrical components often involve copper wires along which electrons flow. Animals, on the other hand, are made of soft, flexible parts, made of flesh, filled with blood vessels and nerves. But animals, as well as machines, need wires and electrical impulses to transmit messages and provide for the control of muscles. But copper wires running through every cubic millimeter of an animal's body would not work. The large physical movements of many parts of the body would quickly break copper wires. If a copper wire were repeatedly bent, back and forth, it would soon break by metal fatigue failure. Do you think you could invent a way to make wires that would not break, and which could transmit electricity, out of flesh? If you don't think you are smart enough to do that, don't fret, it has been done. Meet the neuron.

Figure 5.6 shows a neuron. (Hickman 15) A neuron is a cell, but its elongated central part may be several feet long. Neurons are bundled together into multistrand cables, called nerves, as shown in Figure 5.7. Such a trunk may contain thousands of neurons. The human body contains about 12 billion neurons linked by more than 10 trillion connections. Let's now consider how an electrical impulse travels along a neuron.

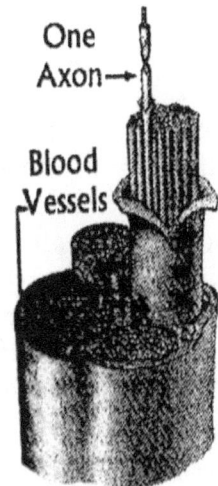

One
Axon→

Blood
Vessels

Nerve Consists
of Bundle of
Many Neurons
Figure 5.7

One way that an impulse can be transmitted along a conductor is by transverse wave action. If a pebble be dropped into a still pond, a wave will be created which may travel across the pond. In such a wave, the particles of water move up and down, but the wave moves horizontally along the water surface. If two people stretch a rope between them and one of them moves his end of the rope up and down, a wave will be created which will move along the rope to the other end. The impulse which travels along a neuron acts similarly, except that the rope or wire consists of a hollow tube, and the substances which move "up and down" actually move inward and outward radially across the membrane which is the wall of the hollow tube. As these substances move inward and outward radially, they create a wave that moves along the neuron.

The substances which do the in-and-out moving are electrically charged particles, ions of sodium and ions of potassium, each having a plus-one positive charge. These ions move through the semi-liquid plasma which exists inside and outside of the neuron tube. The forces which cause the ions to move across the membrane include: (1) electrostatic repulsion and attraction, and (2) the forces of diffusion, under which ions tend to move from a region of high concentration to a region of lower concentration. In addition to these forces, the waves that travel along the neuron are affected by some very remarkable properties of the membrane walls of the tube. First, the walls are filled with thousands of tiny protein pumps that are embedded in the walls. There may be 150,000 of these pumps on each square millimeter of a wall surface. These proteins pump sodium ions from inside the neuron tube to its outside. Also, during the at-rest period, when no impulse is being transmitted, the membrane wall is largely impervious to any movement of sodium ions inward across the membrane, but it permits potassium ions to move across the membrane freely. The forces of diffusion tend to move potassium ions from inside to outside the tube, but the forces of electrostatic repulsion tend to move them from outside to inside, because there are many positively charged sodium ions outside the tube. The net effect of all of these forces causes the at-rest concentrations of these ions to be largely sodium outside and potassium inside, and it produces an electrostatic potential difference across the membrane of about 70 millivolts, with the inside being the more negative.

Now, when a signal is to be transmitted, such as if you put your finger on a hot stove, the membrane, which is the wall of the neuron tube, suddenly becomes about 600 times more permeable to sodium ions, and they rush into the interior. This action triggers the adjacent tube wall to act similarly, and this effect moves along the neuron axially like a wave of water after the dropping of a pebble. After a brief time the at-rest forces again dominate and the sodium ions are moved back outside of the wall, but a wave propagation, involving electrically charged ions, has moved along the neuron.

The inward-outward cycle of wave impulse takes place very quickly, like in one millisecond, and the wave travels along the neuron at a speed that can be as high as 270 miles per hour. Once triggered, the impulse moves quickly to the end of the neuron. All such impulses are of the same intensity. However, to provide for different intensities of signal, the wave action can be repeated, anywhere from a few times per second to as many as 1000 impulses per second, if the stove top is very hot. Also, the intensity of the sensation is related to the number of neurons fired. Finally, neurons operate digitally, rather than by analog action. All triggered impulses are identical. Each wave is either on or off.

Thus, a neuron consists of a hollow tube through the walls of which electrically charged ions move radially to produce a wave action which travels along the tube. This is an electrical phenomenon which acts like electricity moving along a copper wire. But it is all made out of animal flesh, living cells. And it will not break by fatigue even if bent back and forth repeatedly for years. It is an indestructible wire made out of flesh!

Neurons transmit messages to and from the brain, and they cause muscles and organs to perform their various functions.

And now, my dear Reader, let's really think objectively about these neurons. How amazing a feat of design is it that, out of the rocks of the earth, and the water of the sea, and the air above the earth, could be produced a creature having within its body millions of neurons. They work magnificently. How did the tubes get there? How did the sodium ions get there? How did the potassium ions get there? How did the sodium pumps get in the walls? Could you have invented and constructed such a clever device? Do you know anybody who you think is smart enough to invent and fabricate a neuron? Do you think that these wires made out of flesh could have been invented and built by nobody?

There are other examples of electrical devices in animals. The brains of animals are electrochemical devices. The electric eel can produce an electrical potential-difference of 600 volts. And there are many more examples which we will not have space here to reveal.

When I see such complex electrical systems in animals, I see the fingerprints of a very intelligent, very knowledgeable, and very creative supernatural designer and craftsman, who obviously knows more about electricity than I do.

Artistic Patterns, Colors, and Shapes

The exterior surfaces of many animals exhibit patterns and colors which can only be described as works of art. In most cases these decorations obviously have no natural-selection value. They do not hide the animals nor scare away predators. On the contrary, bright colors and geometric patterns would probably do just the opposite, attract predators. Unfortunately, this book is not being produced in color, but most readers have seen the animals to be mentioned and will no doubt agree that many of them are gorgeously beautiful. Space will allow only a few to be mentioned.

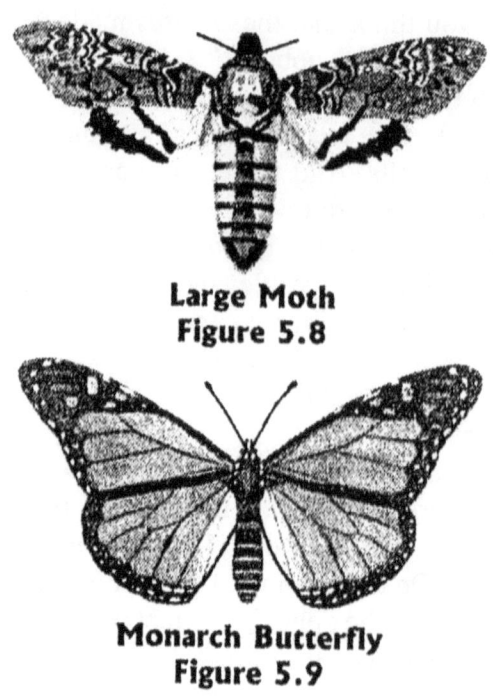

**Large Moth
Figure 5.8**

**Monarch Butterfly
Figure 5.9**

Figure 5.8 shows a large moth, and Figure 5.9 shows a monarch butterfly. Notice in the butterfly the dark black lines, white dots, and variations of orange throughout. It is, indeed, a work of art. The moth contains american-indian-like zig-zag patterns in the fore wings, threaded black bands on a white background on the aft wings, and a hornet-like fuselage. Other animals that show brilliant coloration include birds, fish, coral, frogs, etc. Zebras have distinctive black and white lines, and Dalmatian dogs have black and white spots. Patterns of demarcation and colors are also seen on fruit flies, snakes, pheasants, mollusks, star-fish, turtles, lizards, leopards, and many others.

Some animals have been made into shapes that are, indeed, works of art. Notice the spiral shape of the nautilus in Figure 5.10, (Hickman 16) and the radial symmetry of the star-fish of Figure 5.11. And what person is there alive who would not agree that the body of a young woman who is gorgeously beautiful is not truly a work of art?

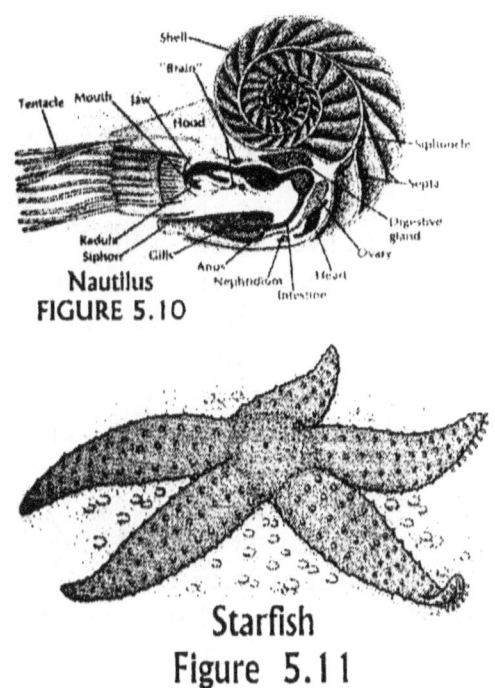

Nautilus
FIGURE 5.10

Starfish
Figure 5.11

Can you explain how a cardinal bird can eat brown seeds produced from brown soil and grow bright red feathers? Or can you explain how an American goldfinch can eat a similar diet and produce bright yellow feathers? Or how can a painted bunting, on a similar diet, develop a blue head, a green mantle, and red underparts? Have you seen some of the birds of Africa?

You know as well as I do that there are hundreds of other examples of pattern, color and shape in the animal kingdom, and that these features have nothing to do with the survival-of-the-fittest, or with natural selection. These are additional fingerprints of an intelligent designer-craftsman, who also enjoys beauty, and who obviously could teach a fascinating course in creating and appreciating art.

Clever, Novel, Patentable Devices

Finally, we can identify the last of the ten footprints and fingerprints which I use to determine that an animal has not evolved, but rather, has been invented and created by a master designer. When I see a device that is a part of an animal, and which is obviously very clever, novel, and which actually, in my professional opinion, would be patentable if it were conceived by a human being, I can logically conclude that that device had to be invented by an intelligent designer.

Each country of the world has a patent office which issues a patent to the designer of each applicant who has invented a truly new, novel, clever, and useful device. For example, Thomas Edison invented the phonograph and the electric light bulb, and the United States Patent Office issued to him a patent for each. I know something about patents and patentability because I have conceived several patentable devices myself, and I have served as the expert witness in several multi-million-dollar patent lawsuits. As I study animals, and consider them as machines, I see many devices that are, indeed, clever, novel, and patentable. To illustrate, let me describe a few.

First, consider the stinging organoids which are carried by many of the sea creatures, such as the Portuguese man-of-war, jellyfish, sea anemones, hydras, and others of the phylum, Cnidaria. Many of these animals have tentacles and body parts which are covered with stinging capsules, which, when activated, shoot out tiny barbs covered with poison. One, known as the sea wasp, is so poisonous that it can cause death to its prey, even human beings, within a few minutes.

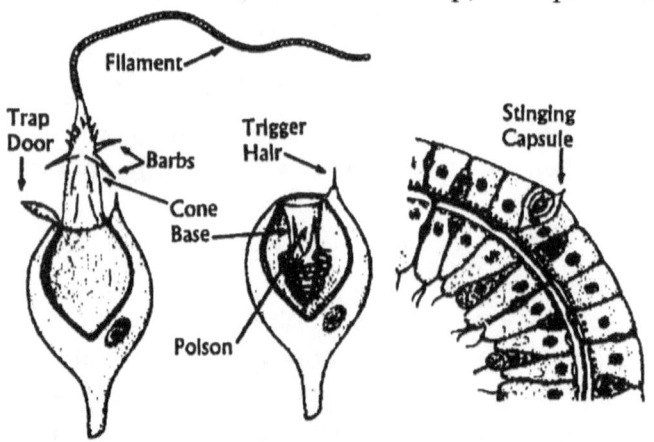

Figure 5.14
Stinging Capsule
After Activation

Figure 5.13
Stinging Capsule.
Before Activation

Figure 5.12
Wall of Tentacle of
Stinging Sea-creature

Figure 5.12 (Hickman 17) shows the outer surface of a tentacle, or body part, of such a creature. Notice the stinging capsule which is grown embedded among the cells of the surface of the part. Figure 5.13 shows an enlarged view of this stinging capsule with its active parts all coiled up in the capsule, before being activated. Figure 5.14 shows the discharged capsule after activation. Before discharge, all of the thread-like filament, the cone-shaped base, and the barbs shown in Figure 5.14 are coiled up inside the capsule, as shown in Figure 5.13. While in the coiled position the insides of the coiled parts are bathed in a poisonous substance. Before discharge, the capsule is sealed by the trap door shown in Figures 5.13 and 5.14. The capsule also contains a trigger hair, shown in both Figures 5.14 and 5.13, which extends outward to sense when an enemy or a prey has gotten too close. Due to the phenomenon of osmosis a large pressure develops within the capsule when it is in the at-rest status. When the trigger device is touched by an enemy, a most remarkable sequence of events takes place. The membrane of the active parts suddenly becomes permeable to water, somewhat like the action of neurons, and the high osmotic pressure within the capsule becomes applied to the inside of the active parts. The trap door flies open and the cone base and all of the stinging filament turn inside out, and these elements fly outward with great force. The active parts actually turn inside out. This causes all of the barbs and the poison to be exposed on what is now the outside of the cone base and its long filament. The enemy is hit, pierced and impregnated with poison, in a fraction of a second. Since the surfaces of the animal's tentacles and body are covered with many such stinging cells, the enemy will get hit by many of these poisonous filaments and barbs.

Now, I don't know what your opinion is, but I consider this stinging device to be extremely clever, certainly novel, useful, and, without question, patentable. If a human being had first invented this device, the Patent Office would have granted him a patent for it. Which of the natural forces of the universe could produce a device like this with no help from any person? Tell me if you know. I would testify in court that this device had to be invented by some intelligent designer.

Among the animals of the earth there are literally hundreds of other examples of clever, patentable devices like this. Let's consider briefly just a few of them. Note the beak of the bald eagle shown in Figure 5.15. (Bender 1) Note how the upper part of the beak turns downward at its outer end, and the lower part fits up into the L-shaped bend of the upper portion. Did you ever try to rip apart a tough steak with a pair of chopsticks? The beaks of most birds, seed eaters, are straight, like chop-sticks. They are not very useful for ripping apart meat. But the eagle eats meat. Its beak is designed to easily rip pieces of flesh out of prey. The bent part of the upper portion sinks into the flesh like a knife, and the lower part then clamps it there while the eagle pulls on both parts and rips out chunks of meat. This beak acts

**Figure 5.15
Eagle**

like a backhoe digging into tough earth. Without the bent end of the upper portion, with the lower beak to provide for clamping, the bird would starve. A bird of prey is well served by having a beak of this very clever design.

Consider next the common house fly. At the end of each of its six legs there is a pair of tiny claws that resemble those of a lobster or a crayfish. With these, the fly can grasp the rough surface of a ceiling and walk upside down. But now, notice that the fly can also walk upside down on the bottom of a sheet of glass, where there are no rough protrusions to which it can clamp its claws. How can it do that? Further investigation will reveal that on the end of each leg, in addition to the claw, there is a suction cup. The fly can squeeze the air out of these cups and attach its legs to the glass by suction. It is literally held up there by the air pressure of the atmosphere on these cups. I think these attachment devices are very clever and novel.

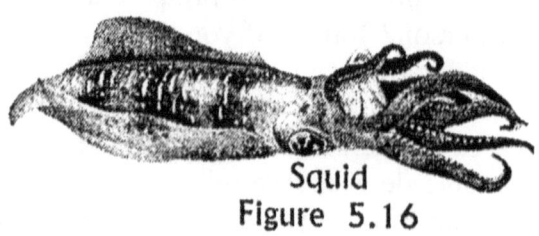

**Squid
Figure 5.16**

Squids and cuttlefish are sea creatures which can propel themselves by the action of a jet-propulsion system. The squid, shown in Figure 5.16, takes water into its mantle cavity as certain muscles relax. Other muscles then can close the neck opening in the mantle and squeeze and propel this water with considerable force out of the funnel, which is located on the underside of the body near the head. The squid, being streamlined in shape, can then propel itself at high speed through the water, using this jet-propulsion system. I think this device is very clever and novel.

The flight of any bird or insect is remarkable, and will be discussed in more detail in Chapter 12. But here, to illustrate clever devices, let's consider the hummingbird. The action of wings in general is not simple. If an airfoil such as the wing of a bird were to be moved quickly downward, the reaction of the air on the wing would tend to lift the bird upward. If the airfoil were then moved quickly upward, the air reaction would tend, with equal force, to move the bird downward. These two effects would cancel each other. Simple up-down movement of wings will have no net lifting capability. Yet the smallest hummingbird, which weighs about as much as a five-cent nickel, can vibrate its wings at 80 times per second, and move at a speed of 60 miles per hour, or hover motionlessly above a flower. It accomplishes these feats by causing the wings to change shape and actually twist during downward motion as compared with upward motion, and the wings follow a motion path which is not just up and down. There are forward and reverse components in the wing motions.

Other very clever devices in the animal kingdom could be cited, including, eyes, ears, the poisonous fangs of snakes, cat's claws, the life cycles of butterflies, the balance system in the inner ear of mammals, the heat sensors of snakes which can detect temperature differences of as little as 0.005 degrees Fahrenheit, the antifreeze, glyceral, which moth pupae generate to keep them from freezing to death in the winter, spider's webs, milk-producing breasts, the electric eel which can produce a shocking voltage, the skunk's defense system, sweat glands, the sonar system for bats, and many, many others.

Do you think that these ingenious, clever, novel, and patentable devices could come into existence by themselves, without the deliberate mental creative efforts of an intelligent designer? What would be the reaction of the Patent Office if you presented them with the plans for one of these devices and told them that the inventor was nobody? Evolutionists believe that such devices came into existence by the chance actions of random natural forces, followed by the effects of natural selection and the survival of the fittest. As I have stated before, I agree that natural selection is a valid force; it causes the unfit to become extinct. But how could random natural forces, alone, bring about such complex, purposeful, ingenious devices as those cited above? The second concept of the theory of evolution is valid, but the first, in my opinion is incredible.

These Fingerprints Tell Who Designed the Animals

The purpose of this Chapter has been to describe for you several of the tell-tale fingerprints I look for when I analyze an animal to look into the mind of its designer and determine its origin. We have considered ten different types of fingerprints. These have included: ordered arrays of materials, well-engineered shapes of parts, the use of refined materials, effective manufacturing processes, the use of multipart systems, complex mechanical systems, complex chemical systems, complex electrical systems, artistic patterns or colors, and clever and novel patentable devices. This Chapter should also give you more in-depth insight into the reasoning processes that anyone should employ to determine, really, who designed and constructed the animals of this earth.

CHAPTER 6. ANIMALS ARE MACHINES

In Chapters 2 and 3, I explained what an expert witness is, and I presented my qualifications for being classified as an expert in my field, which is the design of machinery. I must now prove to you that animals are machines, and then I can apply my expertise to the matter of how animals have been designed, and who did the designing.

Definitions of Machines

We must first determine what is a machine. *The American Heritage Dictionary* states that a machine is, "Any system, usually of rigid bodies, formed and connected to alter, transmit, and direct applied forces in a predetermined manner to accomplish a specific objective, such as the performance of useful work." The *Dictionary* then adds, "Any such system or device, together with its power source and auxiliary equipment, for example, an automobile, aircraft, or jackhammer." Then, believe it or not, the *Dictionary* further adds, that a machine is "An intricate natural system or organism, such as the human body."

We could stop here and say that our case has already been proven by the *Dictionary*. If it says that "an intricate natural system" is a machine, and then cites the human body as an example, surely we can conclude therefrom that animals, which are clearly also "intricate natural systems," must, similarly, be defined as machines. Thus, animals are machines by definition!

As convincing as it may be simply to quote a dictionary definition, we don't want to rely only on semantics in our effort to convince you that animals are machines. We want to consider this matter in much deeper detail by studying substantively what are machines and what are animals. By comparison, then, this should provide final convincing evidence that animals are machines, and a machine designer can analyze them as such.

We all know that definitions can be short or long, depending upon the amount of detail that is included in the definition. A somewhat longer definition, which I composed, states that a machine is an assemblage of strong and rigid parts, and sometimes gaseous or liquid fluids, connected together with joints, bearings, and fasteners, in such a way that some of the parts can move with respect to the others; and, when energy is applied to the system, one or more of the parts will move and/or exert forces which perform useful functions or do useful work.

Examples of Machines

Table 6.1 below lists and categorizes various examples of machines.

TABLE 6.1. EXAMPLES OF MACHINES

GENERAL TYPE OF MACHINERY	FUNCTION TO BE PERFORMED	MACHINERY EXAMPLE
Transportation	To transport freight or human beings	Automobiles Trucks Trains Motorcycles Boats Aircraft Conveyors Elevators
Agriculture	To till the soil, plant crops or harvest crops	Tractors and plows Disks and harrows Seed planters Cultivators Hay-making tools Combines Bulb diggers
Mining	To mine minerals or petroleum	Drilling machinery Rock excavators Rock crushers Rock separators

Energy conversion	To transform energy from one form to another	Diesel engines Gasoline engines Steam turbines Gas turbines Electric motors Generators Steam power plants
Machine tools	To change the size and shape of solid materials, to make machine parts	Lathes Shapers Milling machines Drills Forges
Fluid-flow equipment	To move, constrain, or pressurize gasses or liquids	Pumps and pipe systems Fans and duct systems
Household appliances	To perform various household tasks	Vacuum sweepers Food mixers Washers Dryers Refrigerators Air-conditioners Heaters Mowers
Earth movers	To move earth from one place to another	Bull dozers Back hoes Front loaders Ditch diggers Dredges Carryalls
Equipment of warfare	To provide personal Protection or national defense	Guns Missiles Fighter aircraft Artillery

Animals	To enable self-transportation, food gathering, home building, enjoyment of the environment, and procreation	Land animals Birds Fish Insects Human Beings

All of the above examples are machines. They all have strong and rigid parts which move, and they are activated by some source of energy, such as from the burning of gasoline, diesel-fuel, coal, food, or the consumption of electricity, as by electric motors. And all perform useful functions or do useful work.

Machines Contrasted with the Products of Other Engineers

Machines may be contrasted with most of the products designed by other designers, such as electrical, chemical or civil engineers. In electronic devices, such as television sets or computers, it is the electrons which move, not solid parts. In chemical reactions, it is the atoms which move and align themselves differently, not solid parts. And the civil engineers design structures such as bridges, buildings, and electric-power-transmission towers, which structures have no moving parts. Machines have solid parts that move, and they are designed by mechanical engineers who are machine designers.

Machine Elements

Machinery can be further understood in depth by learning what are the machine elements, or the basic building blocks, of which machines are made. These include such elements as the following: levers, connecting rods, pistons, crankshafts, machine frames, bearings, gears, shafts, brakes, clutches, springs, hydraulic cylinders, pipes, pressure vessels, pumps, belts, pulleys, cams, roller chains, wheels, nuts, bolts, rivets, weldments, chains, cables, ropes, flywheels, fasteners, turbine blades, propellers, airfoils, lubricators, control systems, optical devices, auditory devices, heating, ventilating and air-conditioning equipment, and others.

Complex Machines

Complex machines are made up of combinations of the above-identified machine elements. An automobile, for instance, involves many of the above elements; actually it involves most of them. Furthermore, many machines also involve, usually as sub-systems, products and processes designed by other engineers. An automobile engine involves chemical reactions, the burning of fuels; and electrical devices, such as for ignition. But, the majority of the parts of an automobile are the machine elements of the mechanical engineer, and an automobile involves rigid parts that move, and its function is a mechanical function, ie. to transport people or freight. Therefore it is basically a mechanical device, and it is designed by a machine designer, who is a mechanical engineer. Machine design is a sub-topic under mechanical engineering.

Are Animals Machines?

Are animals machines? Obviously they are! Animals have strong and rigid parts. The bones are fastened together by bearings and joints. The parts move with respect to each other. An animal can transport itself. It can haul freight. It is powered by energy it absorbs. It processes and burns its food. Many animals contain levers, bearings, springs, cylinders, pumps, pipes, pressure vessels, cables, lubricators, control systems, optical devices, auditory devices, and others. The bones and skeletal parts of animals are strong and rigid parts. Limbs, such as arms and legs, move with respect to a frame. Bones are often connected to each other by bearings. The bearings are cushioned by spring materials, and are lubricated by lubricating fluids. The heart is a pump, which pumps blood, a liquid. Blood flows through pipes, the blood vessels. Each lung is also a pump. It pumps air. Body ligaments are really cables. A temperature-control system is present in many animals. In humans, the body temperature stays close to 98.6 degrees Fahrenheit. Eyes are optical devices. Ears are auditory devices. The sweating phenomenon is an air-conditioning system.

Animals perform many of the major mechanical functions, which shows that they are machines. They transport themselves, and they haul freight. They haul building materials from which they build their homes. Birds build nests in trees. Beavers build houses in lakes and ponds. Bears dig holes under trees. Animals are agricultural machines. They harvest crops for their food. They engage in mining. They dig holes and move rocks. They transform energy. Their muscles transform chemical and electrical energy into mechanical work. Their hearts and lungs pump blood and air. Their blood vessels are mechanical pipes. They perform household tasks. They remove garbage and wastes, and they process food for their young. They wash themselves and their young. Many animals are prolific earth movers. Some construct tunnels; others dig burros; some construct mounds. All animals provide for self defense. Some cooperate in group warfare.

Animals are Machines!

Are animals really machines? Indeed they are! In the first place, they are machines by definition. Secondly, it should be apparent from the in-depth study presented above that the characteristics of machines and the distinctive features of animals provide an excellent match. And, finally, I am an expert in the field of machinery, and I say that animals are machines. It is a fact, animals are machines!

CHAPTER 7. DIFFERENCES BETWEEN MAN-MADE MACHINES AND ANIMALS

Dissimilarities are Just Different Design Approaches

Motorcycle

Horse

Although we demonstrated in Chapter 6 that animals are in fact machines, there are some interesting general differences between animals and man-made machines. It should be appreciated, however, that these differences do not in any way diminish the importance of the many similarities between man-made machines and animals, and these differences do not detract from the fact that animals are machines. The dissimilarities are actually just different approaches, or mechanical mechanisms, that were selected by their designers to achieve equivalent objectives. The analogous parts of an animal are, in fact, machine parts just as much as are the counterparts in a man-made machine. For example, animals use reciprocating legs for movement, whereas automobiles and motorcycles use wheels. But a leg is just as much a machine part as is a wheel.

Design of Animals is not Inferior to Design of Machines

The purpose for studying here the differences between animals and man-made machines, is to demonstrate in considerable depth that the various parts of animals are clever and sensible machine elements, and, although the machine systems designed into animals may be different from their machinery counterparts, they clearly are not inferior. Indeed, in many cases, the machine parts and systems of the animals are superior to the comparable elements in man-made machines.

We know who made the man-made machines. Possibly this in-depth study of animal-machines might help us determine who made the animals.

Differences Between Animals and Machines

These differences are outlined in the following Table.

TABLE 7.1
DIFFERENCES BETWEEN MAN-MADE MACHINES AND ANIMALS

MAN-MADE MACHINES	ANIMALS
1. The raw fuel is of fossil origin, mined elsewhere, by separate machinery.	The raw fuel is food gathered by the animal itself.
2. Raw fuel is refined elsewhere by separate machinery to produce gasoline, diesel-fuel, crushed coal, etc.	Raw fuel is refined by the digestive system within the animal itself, to produce refined fuel, such as sugar.
3. Refined fuel is supplied directly to the energy-conversion device, the engine, from a storage tank.	Refined fuel (sugar) is transported by the blood to the energy-conversion devices, the muscles.
4. Air, which includes oxygen, is also supplied to the energy-conversion engine.	Air, which includes oxygen, is pumped into the lungs and it, too, is transported by the blood to the energy-converting muscles.
5. The energy-converting engine then converts the internal energy in the fuel into mechanical work.	The energy-converting muscles then convert the internal energy in the fuel into mechanical work.
6. The energy-conversion is performed at one central location, at the engine.	The energy-conversion is performed at many locations, where needed, at each of the many muscles.

7. Exhaust is removed at one central location, through the exhaust ports of the engine.	Exhaust is removed at many locations, at each of the many muscles. It then goes through the blood back to the lungs, then out of the mouth.
8. The mechanical work produced by the engine must be transported to the location of the ultimate work-performing areas by power-transmission machine elements, such as shafts, gears, belts, pulleys, etc.	The mechanical work produced by the muscles is already at the location of the ultimate work-performing areas, with no need for shafts, gears, belts, pulleys, etc.
9. Many man-made machines have rotary parts.	All animals have reciprocating parts only.
10. Many are high powered.	All are low powered.
11. Many operate at high speed.	All operate at comparatively low speed.
12. All need an operator who has a brain.	All have a built-in brain.
13. None have any instincts.	All have significant instincts.
14. None can self-repair damage.	All can self-repair damage.
15. Most are single-purpose devices.	Most can perform many tasks.
16. None can produce offspring.	All can produce offspring.

These differences are worthy of more detailed discussion.

Sources of Fuel

One of the major differences between man-made machinery and animals has to do with the systems for acquiring and processing fuel, and utilizing the fuel in the devices which transform the internal energy within the fuel into mechanical work. For man-made machines, the fuel is mined, processed, and refined externally to the machine by other machines and processing plants. For example, crude oil is pumped from the earth and it is then processed and refined at a refinery into gasoline or diesel-fuel, all by separate machinery at other locations. Similarly, coal is mined at one location, crushed at another, and it is then transformed into electricity by a separate power plant. The refined energy sources, such as gasoline, diesel-fuel, or electricity, are then supplied to the man-made machines that are to do work.

In the case of animals, the raw source of energy is the food which the animal eats. It is gathered by the animal, it is eaten, and it is then processed by the digestive system within the body of the animal to a refined fuel, usually some form of sugar.

Energy-Conversion Systems

In all machines, man-made or otherwise, an energy-laden fuel must be supplied to some energy-conversion device which converts the energy in the fuel into mechanical work. Obviously the mechanical work ultimately must be performed at the location where the work needs to be performed, such as at the wheels of an automobile or the legs of an animal. At this point, the term "mechanical-work" probably needs to be defined so the reader will have a better understanding of the engineering concepts being developed here. Mechanical-work is the mathematical product of force times the linear distance through which that force acts, or the product of torque times the rotary distance through which the torque acts.

In order to get the energy out of most fuels, the fuel must be combined, chemically, with oxygen, so that it can be burned and expanded to provide useful work. Examples of such energy-conversion devices are internal-combustion engines, gas or diesel engines, such as those which power automobiles and trucks. Or, the combination of a central power plant, which produces electricity, and an electric motor in your home is another example. Coal and air are combined at the power plant to produce electricity. The electricity is then transported to your home, by means of wires, and it can then be directed to a machine, where it drives an electric motor. The central power plant plus your washing machine is an example of this type of energy conversion.

In the case of an animal, the animal gathers food, eats it, and then the digestive system of the animal processes and refines the food into fuel, usually some form of sugar. This sugar is then inserted into the bloodstream. Simultaneously, the lungs of the animal pump in air, which contains oxygen, and the oxygen is also inserted into the bloodstream. Both the fuel and the oxygen are then transported by the blood, by means of hundreds of pipes, the blood vessels, to the various muscles of the body, where the work needs to be performed. The muscles are the energy-conversion devices. They convert the energy in the fuel into mechanical work. The muscles produce the needed forces and motions.

Locations of the Energy-Conversion Devices

In the case of man-made machines, all of the energy conversion is done at one location, by the engine or motor. The output of the engine or motor is mechanical work, usually in the form of torque acting through an angular displacement. This rotary energy is then transmitted by machine elements such as shafts, gears, belts, pulleys, or hydraulic equipment, to some ultimate machine part where the energy is needed, such as at the driving wheels of an automobile, the beaters of a food mixer, or the propeller of an aircraft.

In the case of animals, the fuel and oxygen are transported by the blood directly to the energy-conversion devices, muscles, at which location the work is done directly, with no need for other power-transmission devices. Animals do not need shafts, bearings, gears, belts, etc. The work is done precisely where it is needed, at the muscles. The energy is transported by the blood before it is converted into work.

Exhaust-Removal Systems

Whenever energy is transformed by the burning of a fuel, there are toxic exhaust gasses which must be removed. In the case of engines, these exhaust gasses are removed by running them through a pipe leading away from the centrally located engine. In the case of a central power plant, which is producing electricity, the exhaust gasses are directed up a tall chimney. In the case of animals, the unwanted gasses are inserted at the location of the muscles into the blood stream and through it they are transported to the lungs. The lungs then take the unwanted gasses out of the blood and send them through pipes to the mouth, from which they are exhaled into the atmosphere.

Animal Control Systems

In addition to blood vessels, an animal's body also contains nerves, widely distributed throughout the body. If an animal were to be pricked with a pin at almost any point on its body, nerves would signal this fact to the animal's brain, and blood would probably flow from the wound. The machines made by human beings contain no such networks of nerves and blood vessels. The machines designed by human beings are typically made of solid pieces of metal or plastic. There are no nerves or blood vessels in the fenders of an automobile. We discussed above the functions of the blood and blood vessels in animals. Now we need to learn what are the functions of the nerves. One of the primary functions of the nerves is to send messages from the brain to the muscles to tell each muscle when to convert energy and exert force. The nerves control the actions of the muscles. In order to perform some work task, it usually requires the coordinated efforts of many muscles, all operating at the same time. It is the nerves, together with the brain, which provide these command and coordinating functions. No man-made machine is similarly filled with such nerves and blood vessels.

Animals Operate by Reciprocating Motions

Another interesting difference between these two types of machines is that man-made machines often operate by rotary motion, but all animals operate only by reciprocating motion, never by rotary motion. For example, an internal combustion engine, such as is used in automobiles, produces rotary motion. The output shaft rotates. Turbines produce rotary motion. Most transport vehicles, such as autos, trucks, trains, etc., move on rotating wheels. Animals, on the other hand, move on reciprocating legs. Aircraft are propelled by rotating propellers or rotating jet engines. Birds fly by up-and-down, or back-and-forth, reciprocating wing action. Boats are propelled by rotating propellers. Ducks swim by reciprocating webbed feet. Agricultural combines operate largely by rotating machine parts. Animals harvest food by the reciprocating actions of limbs and teeth. Centrifugal pumps rotate. An animal's heart pumps by in-and-out reciprocating squeezing action. Food mixers have rotating blades. Animals process food by reciprocating jaws and teeth. Most modern ditch diggers are rotary. Ground hogs dig by reciprocating movements of their front feet and claws. Dozens of other examples could be cited.

Advantages of Reciprocating Motions

We might well ask why animals perform all of their functions using reciprocating motion. One reason is that most movements of animal parts are powered by muscles, and muscles operate like hydraulic cylinders. They produce linear motion rather than rotary motion. Also, the bodies of animals are filled with blood vessels and nerves. These would be twisted in two and ripped apart, if body parts had to rotate one with respect to another. Rotary motion is not compatible with parts that are filled with blood vessels and nerves.

Actually, for the important needs of animals, there are advantages that reciprocating designs have as compared with rotary designs. Animals are often required to travel over rough, rocky terrain, or climb steep hills, or move through water or deep mud. How far could one go on such terrain if he were confined to being in a wheel chair? Wheels get stuck in rough or slippery terrain. When life depends on not getting stuck, the reciprocating motions of legs and arms are, by far, the more reliable design. In the case of many other functions that need to be performed, reciprocating motions are the most reliable and hence are the best for animals.

Some man-made machines do operate using reciprocating motions. Examples include the following: back-hoes, front-loaders, shapers, displacement pumps, sickle-bar mowers, bulldozers, dredging machinery, clothes washers, and many others.

Animals do not Need High Power or High Speeds

Another difference between man-made machines and animal-machines is that some man-made machines operate at very high levels of power and speed, whereas all animals operate at low levels of these parameters. For example, a horse can produce mechanical work at the rate of one horse-power, but an automobile engine can easily operate at the power of 200 horses. Aircraft can move at 1500 miles per hour, whereas the fastest birds can only fly at about 50 miles per hour. An elephant can push with a large force, but a bull-dozer can exert several times as much force as an elephant.

The basic reason for these differences in levels of power, force and speed is the fact that human beings have been able to find and use fossil fuels, such as oil and coal, which can be stockpiled and then later burned in enormous quantities to produce energy at very high levels of power. An animal can only use the small amount of energy provided by the food that it can gather at a slow rate with its very limited capability for food gathering and food storage. It cannot stockpile huge quantities of energy.

One of the primary reasons that much of our man-made machinery is rotary, is the fact that rotary machinery is more suitable for high-powered and high-speed applications. If parts must move back and forth rapidly, undesirable dynamic forces are induced. Reciprocating motions are very difficult to balance, dynamically, but rotating machinery is very easy to balance. Rotary machinery is very suitable for operating at high speeds. A jet engine of an aircraft can easily operate at very high speeds and very high levels of power. Such an engine can produce an extremely large propelling force. But animals do not need to operate at high speeds or at high levels of power. All they need to do is to move so they can gather food and enjoy life.

Animal's Brain Serves as its Operator

All man-made machines require an operator. Machines are made with control levers, switches, knobs, etc., and it is necessary to add to the machine a human being, who has a brain, to operate the machine. An automobile needs a driver. An agricultural combine needs a farmer who knows what he is doing. Some machines are designed with built-in controls which can provide some operation, but it is necessary, at least, to have a human being to adjust its controls and turn it on and off at the proper time.

Animals, on the other hand, have a built-in brain that makes decisions and provides operational control, and no additional operator is required.

Animals are Guided by Instincts

Another difference has to do with instincts. All animals seem to have a built-in ability to take certain actions in response to particular stimuli. They appear to be pre-programmed to do certain things, or respond in certain ways to situations that arise. For example, newly-born animal babies clearly seek out the mother's nipple and suckle on her breast to get food. Each animal seeks out the particular food for which its digestive system is suited. A quick movement of an object toward an eye causes the eye to close in self defense, or blink. Each animal seems to know how to fertilize the eggs of the female of its species. Birds build nests, beavers build dams, predators hunt, cows eat grass, all by instinct.

Man-made machines have no such instincts. It is possible for human beings to design and build into a machine the ability to respond in certain ways to selected stimuli. For example, a military aircraft can be programmed to follow the terrain and fly at a specified height above the ground. A machine tool can be made to follow a given built-in program. But these actions are not usually referred to as instincts. They are very limited in capability.

Many animals, especially insects, seem to operate for a high percentage of their total activities in response to instinctive direction. Ants and bees build nests, gather food, fight wars, and do many other remarkable things, solely guided by complex instincts.

Animals have Automatic Self-Repair Capability

Another difference is the ability that all animals have to self-repair damage to their bodies. The size and shape, and all of the millions of details, of the body of an animal seems to be under the control of its DNA. Somehow this controlling agent monitors the body, and, if it is damaged, the body, by itself, repairs the damage. If I cut my finger, miraculously, it heals itself.

Man-made machines do not have this capability of self-repair. If I dent the fender of my automobile, it has no capability to repair itself. I must take it to a body shop to have it repaired.

Animals can Perform Many Tasks

Most man-made machines tend to be designed to accomplish only one single purpose. An automobile transports people. A lathe produces round parts. A lawn mower mows lawns. A ditch-digger digs ditches. A corn-planter plants corn. A vacuum cleaner cleans floors. A coal mining machine digs coal. This single-purpose usefulness is largely because man-made machines tend to have only a single engine, or motor, and a single, or small number of, final work-producing parts.

Animals, on the other hand, have hundreds of muscles, which are work-producing. These muscles all work together to enable an animal to perform an almost infinite variety of tasks. Football or basketball players exhibit marvelous muscle coordination which enables them to perform athletic feats which are most awesome to behold. An animal can easily travel over terrain of almost any degree of roughness. It can struggle through mud, it can climb steep slopes, and perform a wide variety of other feats. It can also gather food and refine it into an ultimate fuel. It has a brain which can make decisions. It can self-repair damage to its body. It is a very, very complex creature, capable of performing many tasks.

Animals can Produce Offspring

Finally, all animals have the remarkable ability to produce offspring. This capability should not be taken lightly. Man-made machines cannot do this. Animals have sex organs, egg-producing organs, organs for egg fertilization, organs for nurturing young, etc. Such organs are of exquisite design and have almost unbelievable complexity.

Man-made machines do not have the ability to produce and nurture offspring.

Animals are More Complex than Man-made Machines

All of the above-described differences between man-made machines and the machines we call animals should serve to acquaint you with some of the general characteristics of machines, and it should inform you of some of the differences between animals and man-made machines. It should also further convince you that animals are machines. Many man-made machines can operate at very high levels of power and speed. But does this mean that man-made machines are more sophisticated than animals, or more complex than animals? Quite the contrary, it should be abundantly clear that animals are much the more complex. Man-made machines have no complex system of nerves and blood vessels, no built-in fuel-refining capabilities, no brain, no instincts, no self-repair capability, and they cannot produce offspring.

All Machines must be Designed and Constructed

Do you think that a man-made machine could come into existence without any designer and without any builder? No complex machine of our modern society could come into existence without the services of an intelligent and skilled designer and an expert builder.

Do you think that animals could come into existence without the services of any designer or any builder? If you believe that, then you probably also believe that you can make a spider out of a rock.

The Cambrian Explosion

CHAPTER 8. DESIGN PRINCIPLES OBSERVED IN ANIMALS.

Natural Forces have Carved the Earth

Here we are, in our vast universe, in our galaxy, in our solar system, on our beautiful earth. As we look about us we can see a wide variety of things that have, somehow, come into existence. Let's consider several categories of these things that we can see on our earth. First, we can see rocks, sand, soil, oceans, lakes, rivers, streams, waterfalls, swamps, glaciers, the sky, clouds, and rainbows. We can observe carved shapes on the earth, such as the Grand Canyon, Mammoth Cave, petrified forests, Bryce Canyon's carved monoliths, natural bridges, natural arches, sand dunes, etc. We can study and determine the laws of physics, chemistry, and mathematics. These laws govern the forces of nature. Without debating the origin of the earth, or the origin of the natural laws, we all agree that these laws exist, and we know what they are. And we all agree that the above-identified structures and features of the earth's surface have been molded by these natural forces.

Animals Can Build Structures

Secondly, we see things about us which have been made by animals. These are primarily structures for gathering or trapping food, providing protection from predators, providing shelter from the weather; and nests for incubating, nursing and nurturing offspring. Birds build nests on the ground or in trees or shrubs. Spiders build webs to trap food. Rodents dig holes and tunnels in the ground. Beavers dam streams, build lakes, and produce houses. Ants and bees build elaborate multifamily cities. We all agree that these structures have been built by the animals, themselves, not by the natural forces alone, and not by human beings. Actually, the designing and building skills of these animals seem to be programmed into them in the form of instincts. The question as to who designed and implanted into them these instincts is another matter that will be discussed later. But we all agree now that these structures are built by the animals, because we can actually witness them performing these constructions.

Humans Design and Build Products

Thirdly, we can see all of the machines and other products, made by human beings. We can observe and study such things as automobiles, aircraft, agricultural machinery, instruments, household appliances, and all of the other machines mentioned earlier which have been designed by mechanical engineers. We can also take note of the structures, electronic devices, and chemical products designed by other engineers. We all agree that these products were designed and built by human beings. We know this because we can actually observe them being designed and built. We may, ourselves, be involved in some of this design and construction, or, we may have friends or acquaintances who are so employed.

Who Designed and Built our Antique Machines?

Next let's consider a corollary of the above, antique or ancient machinery. I enjoy finding and studying antique machinery. As I hike in the Rocky Mountains I often come upon old abandoned mines, and rock-crushing mills. Here I can observe machinery which is very old. I assume that this machinery was designed and built by human beings, but how do I know this? Maybe this machinery was designed and built by nobody. Maybe it was produced by the natural forces of the earth, following the laws of physics, chemistry and mathematics. Maybe iron from the earth, somehow, became separated from its ore, and, by the natural forces of the universe, this machinery became fabricated. I never met the people who might have designed this machinery. I did not see them build it. I could conclude that they don't exist, and that they never did exist. If I made this assumption I would be forced to conclude that the machinery was designed and created by natural forces, alone. I could conclude that these antique machines were made by nobody.

But I have come to a different conclusion. I can see applications of the principles of machine design in these machines. I can imagine what went through the minds of the designers as they designed these machines. I have concluded that, even though I did not see these machines being built, I believe they were designed and constructed by some intelligent beings who were skilled in the design and construction of such machinery. I believe that these builders existed, merely by observing what they built. Also, I am an expert in machinery, and I can see in these machines the obvious applications of many principles of design with which I am familiar. The machines are very clever. They are complex. They are skillfully constructed. I am convinced that these designers did exist. And I think they were human beings. And these conclusions have been arrived at totally by inference, based on studying the machines.

Who Designed and Built the Animals?

And now, finally, let's consider animals. Animals are ancient also. They have been on earth for millions of years. No one has ever seen one created from the raw materials of the earth. There are no living witnesses to their original creation. They are, indeed, antique machines, similar in many respects to the mining machinery mentioned above. But, I believe that I can come to some reasonable conclusions regarding their origin. The basic question of this Chapter is, how can we determine what has been the origin of a machine, and we are including animals as machines. The question is, what must be the reasoning processes by which we can judge what was the origin of a machine, or an animal?

The Basic Ingredients of Machine Design

To answer this question, let's first consider the basic ingredients of machine design. What is the thinking process that a designer must go through as he conceives and develops the design of a machine? The following list includes these major steps:

1. The overall purpose for the machine must first be defined.

For example, an automobile is to transport people. A corn-picker is to pick corn. A ferris-wheel is to give people a thrill by taking them to a high elevation. A forging press is to change the shape of metal.

2. An overall plan, or mechanism must be conceived for the entire machine. For example, for an automobile, an engine driven by steam, gasoline, or electricity, might be selected as the source of mechanical energy. The position of the engine, front or rear, must be selected. Should it be front-wheel or rear-wheel drive? The type of transmission needs to be determined. Gear ratios must be established.

3. Sub-systems must be identified. In an automobile, the engine is a sub-system. The transmission is another. The rear axle system is another.

4. The choice of a mechanism for each sub-system must be chosen. For example, the wing-flap-actuator system on an aircraft could be hydraulic, mechanical, or electrical. A boat could be propelled by jet action, screw-type propellers, or by paddle-wheels. A tractor could have wheels and tires or have metal tracks.

5. The general shape of each part must be conceived.

6. The material to be used for each part must be selected.

7. Detailed engineering analyses and calculations must be performed to determine the final size and shape for each part.

8. Consideration must be given for how each part is to be manufactured.

9. Consideration should be given to final finishes, aesthetics, and appearance.

10. For all man-made machines, a process of experimental development is then initiated. Prototypes are made and tested, and design improvements are implemented.

Machine-Design Principles

As the designer of a machine makes selections, calculations and decisions which determine the sizes and shapes of the various parts, he should apply sound design principles which have been developed over many years by machine designers of the past. Several of these design principles will now be identified and illustrated by examples, in Table 8.1.

TABLE 8.1

MACHINE-DESIGN PRINCIPLES, WITH EXAMPLES

MACHINE-DESIGN PRINCIPLE	APPLICATIONS IN MAN-MADE MACHINES	APPLICATIONS IN ANIMALS
1. The machine-part shape which is the strongest against a wide variety of loads is a hollow tube.	Drive shaft of an automobile. Shafts in aircraft gear boxes. Flag poles. Street-light poles.	Many bones of birds are hollow tubes, as are the arms and legs of many animals.

Hollow Drive Shaft of an Automobile

Hollow Bone of a Bird

2. The strongest multipart structure in a plane is the triangle; the strongest such structure in space is the tetrahedron.	Radar support structures. Electric power transmission towers. Frame structures in agricultural machinery. Roll bars.	The bills of birds are tetrahedrons.

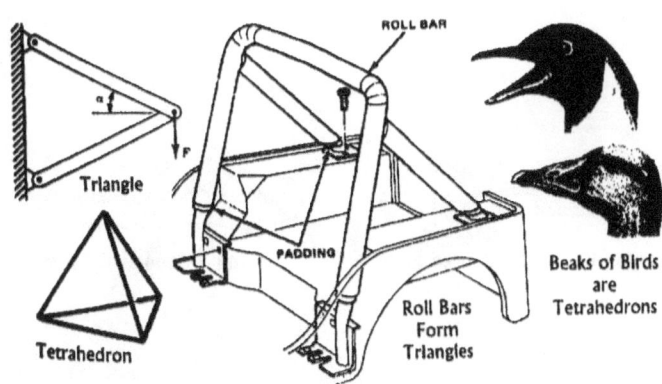

Triangle

Tetrahedron

ROLL BAR

PADDING

Roll Bars Form Triangles

Beaks of Birds are Tetrahedrons

3. The best mechanism for moving parts is the four-bar mechanism and its corollary, the slider-crank mechanism.

The piston-connecting-rod-crank-shaft mechanism of an automobile engine is the slider-crank mechanism, where the piston is the slider. A windshield wiper is a four-bar mechanism.

All arms and legs of animals use the slider-crank mechanism, where the muscles contain the sliding elements.

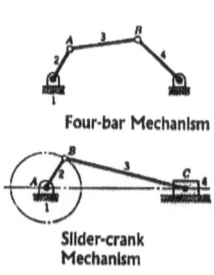

Four-bar Mechanism

Slider-crank Mechanism

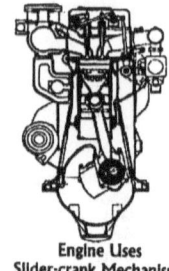

Engine Uses
Slider-crank Mechanism

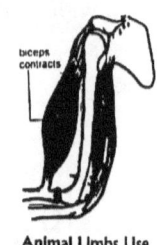

Animal Limbs Use
Slider-crank Mechanism

4. Wherever there is a joint at which two bars are connected to each other for relative movements, the bars must be enlarged near the joint.

Both of the ends of the connecting-rod of an internal combustion engine are enlarged.

All arm and leg bones of animals have enlarged cross-sections at their ends, where the joints are located.

Connecting Rod has Enlarged Ends

Animal Bone has Enlarged Ends

5. Suspension systems which employ spring action give a softer ride and greatly reduce the stresses in all of the parts.

Automobile suspension systems have coil or leaf springs. Motorcycles also have springs.

Animals have knees and hip-joints which, with the associated muscles, give spring action to support the body.

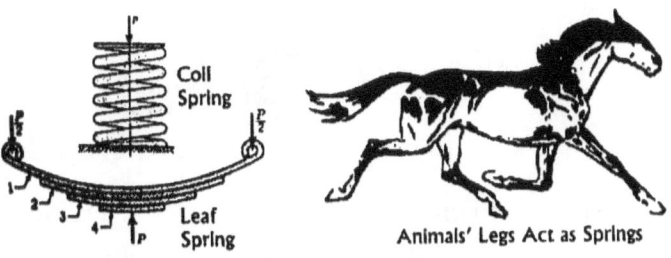

Animals' Legs Act as Springs

6. Where two parts are to grind, crush, or shear other materials placed between them, these parts must be made of very had materials.

Rock crushers have hard liners which act on the rocks which are to be crushed. Metal shears are made of very hard materials.

The teeth of all animals are made of materials that are harder than any other materials of their bodies.

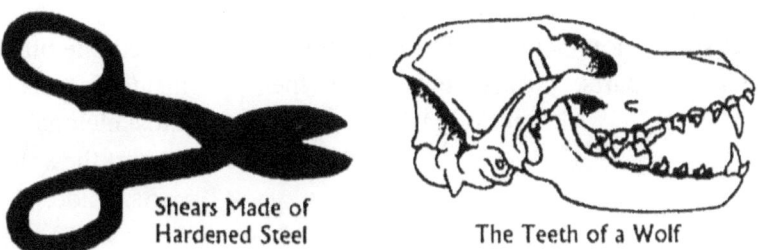

Shears Made of Hardened Steel

The Teeth of a Wolf

7. Where two parts have their relative motion constrained by a bearing, liners made of a softer material should be used.

Soft babbitt or bronze is used to line the bearings of the ends of the connecting rods in engines.

There is a softer cartilage material between the bones of the joints of many animals.

Bearings Need Soft Liners

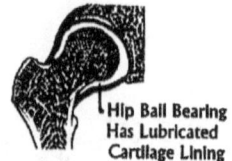

Hip Ball Bearing Has Lubricated Cartilage Lining

8. In bearings where there is sliding motion, a lubricant is desirable.

Engine oil is used in internal combustion engines to lubricate all of the parts of the engine.

A lubricant is provided at many joints between bones in animals. Eye-balls are also lubricated.

9. Shearing action is the best mechanism for cutting other materials.

Steel plates and sheets are usually cut with power shears.

The front teeth of most animals operate by shearing action.

10. Strong and hard housings are useful to protect softer more delicate interior parts.

Automobile bodies protect the much softer people inside. The housings of instruments protect the interior parts.

The head bones of many animals protect the brains inside. Insects have exoskeletons to protect the softer parts inside.

Body of Car Protects Contents

Protective Skull of Brown Bear

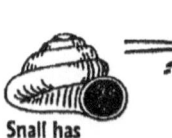

Snail has Protective Shell

Prawn has Exoskeleton

11. Delicate parts can be inset within the surface of a protective housing to minimize damage to the parts.

The door handles of an automobile are inset within the body of the car, to protect them from damage.

Eyeballs are inset within the skulls of many animals to protect them from being hit and damaged.

Door Handle Is Inset In Auto Body

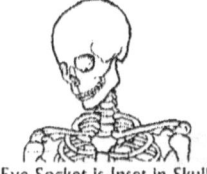

Eye Socket Is Inset In Skull

12. The combined action of many small fasteners, or applicators of force, can add up to a very large total force, and this is better than the use of one large part to develop a large force.

The use of many nuts and bolts, rather than a few large bolts, is widely used in machinery. A zipper is an excellent example of a large closing force produced by many small fasteners.

The human hand has five fingers. Many small muscles are used to develop a large total force. Some insects have many legs. A muscle consists of many small muscle fibers.

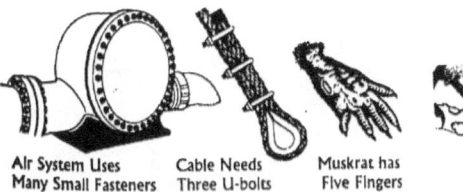

Air System Uses Many Small Fasteners   **Cable Needs Three U-bolts**   **Muskrat has Five Fingers**   **Bird has Four Toes**

13. Structures which provide many small holes where air can be trapped provide the best insulators against heat transfer.

Fiberglas and foam plastic materials are excellent examples of this principle.

The hairy coats of animals and the feathers of birds provide excellent insulators against the transmission of heat or cold.

14. A container, the volume of which cyclically varies, plus two valves, constitutes an efficient device for pumping and moving fluids.

The fuel pump of an automobile. The bilge pump of a boat.

The heart of many animals is a direct application of this design principle.

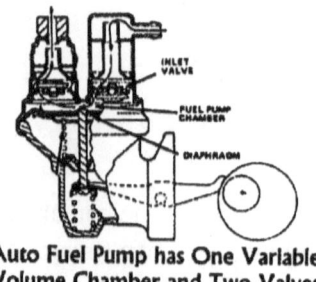

Auto Fuel Pump has One Variable-Volume Chamber and Two Valves

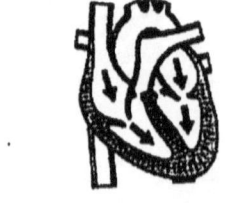

Animal Heart has Two Variable-Volume Chambers and Four Valves

15. The strongest shape for a cantilevered air-foil is one which is triangular in shape and which tapers outward to a point.

The wings of aircraft are of this shape.

This is precisely the shape of the wings of many birds.

Aircraft has Triangular-shaped, Tapered, Cantilevered Wings

Bird has Triangular-Shaped, Tapered Cantilevered Wings

16. The most efficient shape for an object having motion relative to a fluid, to minimize the drag force, is the highly streamlined shape.

Submarines have this shape. The bodies of aircraft have a highly streamlined shape.

Fish are remarkably streamlined in shape.

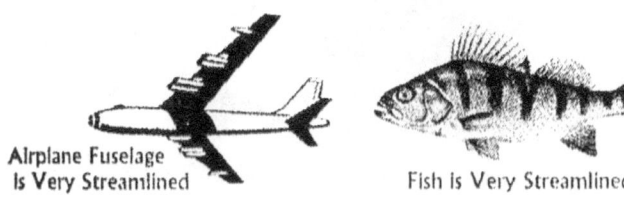

Airplane Fuselage
Is Very Streamlined

Fish Is Very Streamlined

17. The most efficient system for distributing a liquid consists of one large central container followed by many branching pipes of ever smaller size.

The water-distribution system of a city is an an application which follows this principle. A sewer system is the reverse of this concept.

This is precisely the design used to distribute the arterial blood throughout the bodies of many animals. The venous systems are the reverse of this.

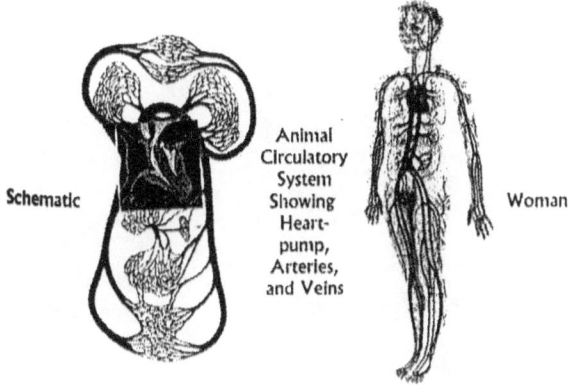

Schematic

Animal Circulatory System Showing Heart-pump, Arteries, and Veins

Woman

18. The most efficient shape for a solid cantilevered limb is a round, gradually-tapered cone.

This is the shape of radio antennas used on ships, and auto gear-shift levers.

This is the shape used for for the tails of most animals, animal horns, and bird beaks.

Gear Shift Lever

Tusks of Walrus

Beak of Great Blue Heron

19. Where one part must be constrained with respect to another part at a single point, but the parts must have relative rotary motion in any plane, a ball-and-socket joint is the best.

A ball joint is used in the front suspension system in many automobiles. It is also the design used for trailer hitches.

A ball-and-socket joint is used at the hip joint in many animals. Eyes also use a ball-and-socket joint in the head.

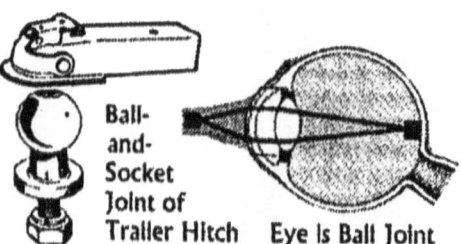

Ball-and-Socket Joint of Trailer Hitch   Eye Is Ball Joint

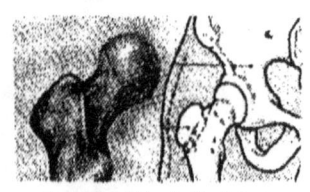

Animal Hip Is Ball Joint

20. Funnel-shaped megaphones are the best means of mechanically magnifying sound.

Listening devices used by armed services are based on this design principle.

This is precisely the system used for the ears of animals.

Funnel-shaped Loud Hailer

Funnel-shaped Ears of Fox

21. An excellent cooling system consists of wetting a surface with a thin layer of water, and then evaporating this water by causing the flow of air over it.

This concept is used in the evaporative cooling systems used in the hot, dry states of the USA.

Animals are cooled in hot weather by sweating and letting the movement of air over the wetted skin or hair cool by evaporation.

Design Principles can be Applied to Analyze Animals

Since I am an expert in the field of machine design, I know the principles of design. I know what are the best shapes to be used for the various applications. I have probably done more research on this subject than anyone in the world. It was I who originated graduate courses in the United States on the subject of mechanical form synthesis. This is the subject of how, scientifically, to select and synthesize the optimum shape that a machine part should have to resist a given set of forces. A graduate student of mine took this material, and, with my permission, wrote a book on this subject. (Marshek 1) This book, and this subject-matter, are new and novel.

Based on this expertise, I can look at and study a machine, and determine the thought processes that must have gone through the mind of the designer as he designed that machine. And, similarly, I can look at an animal, and I can determine the principles of machine design which were applied in order to conceive and develop the designs of the various parts and systems of that animal. I can reconstruct the thought processes that must have occupied the mind of the designer as he designed that animal. I can see design principles applied in the animal's body. And, since mathematics is an important science with which an engineer and a machine designer is familiar, I can calculate the statistical probability that a given animal was created by nobody; that is, by nothing other than the natural forces of the universe.

Can Structures, Machines and Animals be Designed by Nobody?

In Chapter 9 of this book I will perform analyses and calculations that show that there is only one chance in

$$8,320,000,000,000,000,000,000,000,000,000,000,000$$

that a fire ring in the mountains could have been designed and constructed by natural forces alone, and in Chapter 10, I will show that there is even less chance that a single-celled protozoa could have been designed and built by nobody. In some of the later Chapters of this book we will study other animals, and animal sub-systems, and we will point out applications of the above-described general design principles; and I will give my expert testimony as to what must have been the origin of these animal-machines.

Do you believe that animals were made by nobody? If so, please read on, and give honest consideration to the reasoning presented in the remainder of this book.

CHAPTER 9. WHO BUILT THE FIRE-RING?

A Circle of Rocks on a Mountain Top

I have spent many hours hiking and Jeeping in the Rocky Mountains. While on one such excursion, after hiking up a rounded hill, I came to a flat meadow on the top of the hill. There I came upon a group of rocks arranged in a circle, similar to the ring of rocks shown in Figure 9.1. Each rock was about 9" to 10" in diameter, there were 10 rocks in the ring, and the diameter of the ring of rocks was about 30". Other rocks of various sizes were scattered about in a typical random fashion. I wondered how this circle of rocks came into existence in this wilderness area.

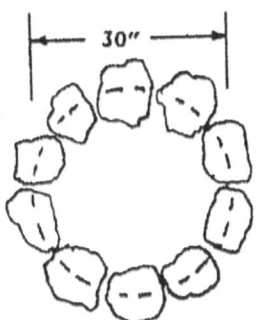

FIGURE 9.1. RING OF ROCKS

Theories on the Origin of the Rock Circle

Two thoughts came to mind. These might be referred to as theories. The first theory assumes that human beings selected a certain size rock, carried some rocks of that size to the location of the ring, and then arranged them into the form of a circular ring, probably to construct a fire-ring, in which they could build a camp-fire.

Under the second theory, I assumed that a group of rocks, all of about 9.4" in diameter, somehow became selected and were transported from the surrounding area to a central location, and they were then arranged into a circle, all by natural forces. Such natural forces might include wind, rain, hail, snow, flooding, freezing, thawing, earthquake, lightning, ice movement, cosmic rays, radiation from the sun, etc.

Which of these two theories do you think represents the truth? Were those rocks put there in that ring by natural forces, or by human beings?

More on the Two Theories

Following the first theory, we might assume that the rocks were put there by human beings! We might assume that one or more intelligent, creative, human beings conceived a design, and constructed that ring of rocks, to accomplish a purpose, namely, to make a fire-ring in which they could build a camp-fire. They obviously picked up rocks from the area around the ring, carried them to a central location, and then arranged them into a circle to form a fire-ring. This involved a purpose, a design, and some construction.

Alternatively, we might assume that the rocks became arranged into that fire-ring configuration by nothing other than the forces of nature. Do you have any idea how improbable it would be for natural forces to produce a fire-ring? Let's explore that possibility. Let's calculate, mathematically, the statistical probability that our fire-ring could have come into existence by natural forces alone, without the assistance of any designer or craftsman.

Natural Forces Must Perform Three Functions

To calculate this statistical probability, we must make some assumptions, and obviously the results of the analysis will depend on the accuracy of the assumptions. We will make every effort to make reasonable assumptions. However, it should be appreciated that only order-of-magnitude results are to be expected from this analysis. But, really, great accuracy is not needed here, and it will soon be apparent that the assumptions we are about to make will be quite adequate to enable us to arrive at some very definite conclusions.

It will be assumed that the natural forces of the earth, on a flat region near the top of a hill, will produce rocks of various sizes, distributed in a random fashion on or near the surface of the earth. The natural forces of the earth must then perform the following three functions: (1) select rocks of about 9.4" in diameter from an area near enough to the final location of the fire-ring to have some chance of being moved there by natural forces, (2) transport some of those rocks to the fire-ring location, and (3) arrange them into a circle of about 30" in diameter. You should remember here that the natural forces of the earth consist of such agents as wind, rain, hail, flooding, freezing, thawing, earthquake, lightning, ice movement, etc., all following the laws of physics, chemistry, and mathematics.

Array from which Rocks may be Selected

We will next make several specific assumptions that will be needed to proceed with our mathematical analysis. First, let's assume that about one tenth of the rocks in the area are of a suitable size, about 9" to 10" in diameter. Second, let's assume that 50 feet is about the maximum distance that natural forces could move a 9.4" rock from its natural position in a random distribution to any other unusual position, such as to the fire-ring center. The time period for our assumptions will be from the time such rocks appeared on this mountain top, to the present.

We can now describe the area from which the rocks could be moved to build the fire-ring. It will be a circular area having a radius of 50 feet. In this area we will assume that the rocks are randomly distributed; that is, each rock is approximately equidistant from each of the other rocks nearest to it. These assumptions will produce an array of rocks, containing 60 to 70 rocks, as shown in Figure 9.2.

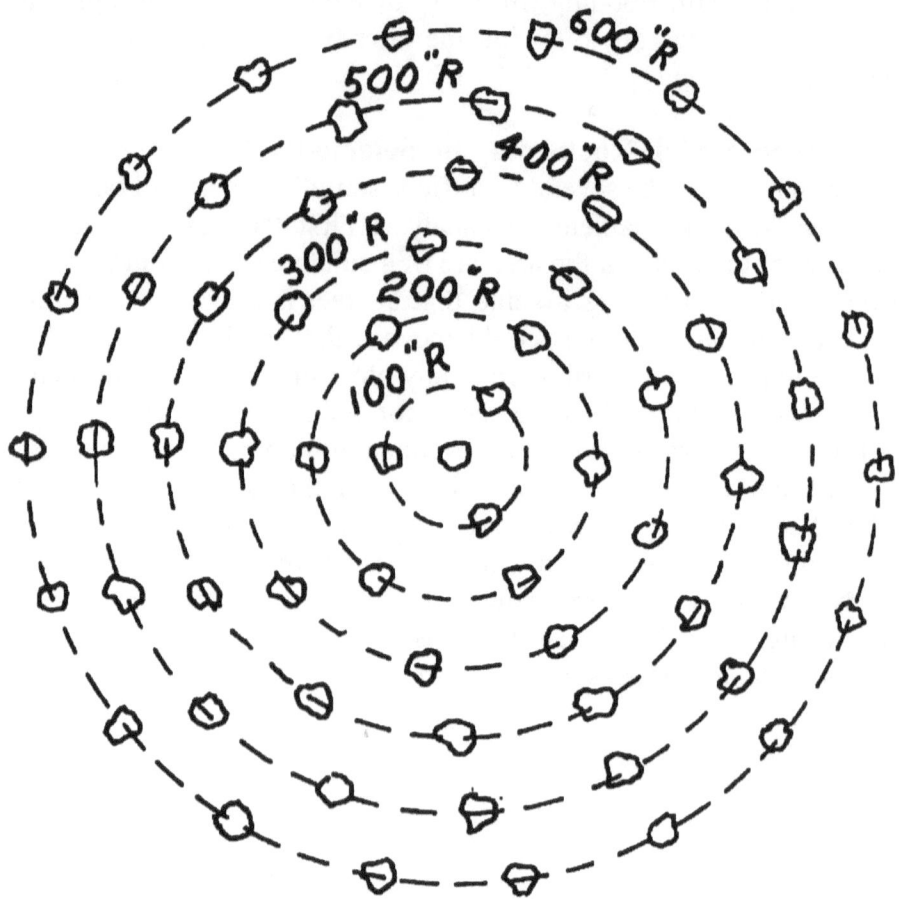

FIGURE 9.2. RANDOMLY-DISTRIBUTED ARRAY OF 64 ROCKS OF 9" TO 10" DIAMETER

To simplify our calculations, we will assume that the rocks in this array are arranged in circles, and each circle will have in it a different number of rocks, so that a uniform distribution can be achieved. The number of rocks in each circle is given in Table 9.1.

TABLE 9.1. DISTRIBUTION OF ROCKS IN SELECTION AREA

RADIUS OF CIRCLE	NUMBER OF ROCKS
0	1
100"	3
200"	6
300"	9
400"	12
500"	15
600"	18

The numbers of rocks at various radial distances shown in Table 9.1 will produce the uniform distribution shown in Figure 9.2. This is the area from which natural forces must select rocks and move them to the center of the fire-ring.

Engineering Assumptions Concerning Rock Movements

We must now make additional assumptions relative to the probabilities that natural forces could move rocks to the central location. The probabilities in this analysis will be assumed to relate to the time period from the present back to when the Rocky Mountains arose and became eroded into small rocks on the surface. Consider any one selected rock that will be moved to the center of the fire-ring. To end up at the center of the fire-ring this rock must not only move the right distance, but it must also move in exactly the right direction. Generally, a rock might move in any direction, but this rock must move in a specific direction, toward the fire-ring.

Probabilities Concerning Distances the Rocks Move

The distance that any rock must move will, of course, be equal to the radius of the circle in the array of Figure 9.2 upon which it is located. Let's assume that there would be one chance in 10 that natural forces could move a rock a net distance of 10" from its natural location. This would appear to be a reasonable, and actually conservative, assumption. Following this assumed probability, there would be one chance in 100 that a rock could be moved 100". This is about 8 feet. Our general assumption here is that the probability that a rock could be moved from one place to another is the reciprocal of the number of inches in distance that it needs to be moved. Based on this assumption, Table 9.2 lists the probabilities that each of the various rocks in our array could be moved the distance required so that it might end up at the fire-ring.

TABLE 9.2. PROBABILITIES OF ROCK DISPLACEMENTS

DISTANCE ROCK MUST BE MOVED, INCHES	RADIAL DISTANCE OF ROCK FROM THE CENTER OF THE FIRE-RING, INCHES.	ASSUMED PROBABILITY THAT SELECTED ROCK COULD BE MOVED THE REQUIRED DISTANCE BY NATURAL FORCES.
0	0	1 CHANCE IN 1
100	100	1 CHANCE IN 100
200	200	1 CHANCE IN 200
300	300	1 CHANCE IN 300
400	400	1 CHANCE IN 400
500	500	1 CHANCE IN 500
600	600	1 CHANCE IN 600

Probabilities Concerning the Directions the Rocks Move

Now let's determine the probabilities involved in having the rock move in the right direction. Figure 9.3 shows the fire-ring, the circles in which the rocks of our array are located, and it depicts a particular rock that must be moved to the center of the fire-ring. Let's call it rock A. Let's assume that rock A is in the array circle that has a radius of 500". Then rock A must move a distance, D, of 500". But, also, it must move in the right direction, toward the center of the fire-ring. The circle, K, has been drawn around rock A to show the various places to which rock A could be moved, if it went the right distance, but in the wrong direction.

To arrive at the probability that rock A would go in the right direction, we need only to calculate how many rocks of 9.4" diameter could be placed on circle K. Our probability would be the reciprocal of that number. For example, the number of 9.4"-diameter rocks that would fit in a circle having a radius of 500" would be, N = (2) (500)(3.1416)/

(9.4) = 334. Hence, the probability that rock A would go in the right direction is, P = 1/334. This means that there is one chance in 334 that the rock would move in the right direction and terminate this part of its journey at the center of the fire-ring. Table 9.3 lists the probabilities that the other rocks in the various circles of our array of Figure 9.2 would move in the right direction. The column headings in Table 9.3, to be accurate and descriptive, need to be somewhat lengthy. So we have used abbreviated headings in the Table.

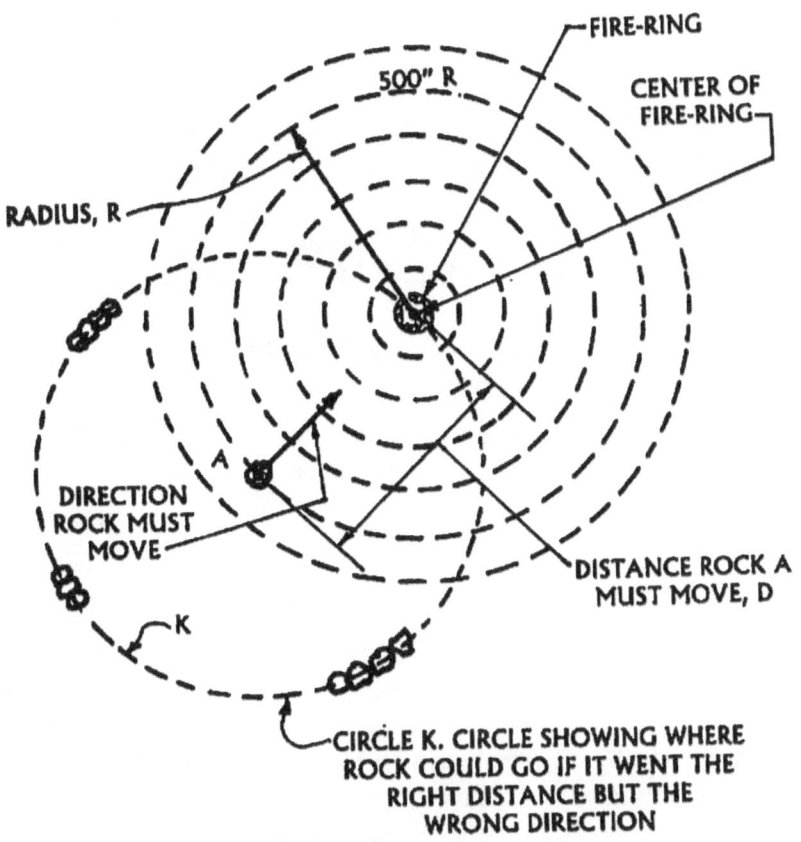

FIGURE 9.3. SKETCH OF A TYPICAL ROCK SHOWING THE DIRECTION IN WHICH IT MUST MOVE

The accurate headings are as follows:

Distance D = the radial distance of a selected rock from the center of the fire-ring, inches.

Circumference K = the circumference of a circle which has the selected rock at its center, and which circle passes through the center of the fire-ring, inches.

Number N = the number of 9.4"-diameter rocks that would fit in the circle, K.

Probability P = the statistical probability that the selected rock would happen to aim in just the right direction to terminate its move at the center of the fire-ring.

TABLE 9.3. PROBABILITIES ROCKS MOVE IN RIGHT DIRECTION

DISTANCE D	CIRCUMFERENCE K	NUMBER N	PROBABILITY P
0	0	1	1 CHANCE IN 1
100	628	67	1 CHANCE IN 67
200	1257	134	1 CHANCE IN 134
300	1885	200	1 CHANCE IN 200
400	2513	267	1 CHANCE IN 267
500	3142	334	1 CHANCE IN 334
600	3770	401	1 CHANCE IN 401

Summary on Rock Movements to the Center of the Ring

So far in this analysis we have done the following: (1) identified an area from which rocks could be selected, (2) assumed that one rock in 10 would be of a suitable size, (3) determined the probabilities associated with rocks moving the right distance to get to the center of the fire-ring, and (4) established probabilities related to these rocks moving in the right direction. It remains now to get the rocks moved from the center of the fire-ring circle to the fire-ring itself, to create the final design.

Final Outward Movements to Form the Fire-Ring

The probabilities associated with this final phase, actually building the fire-ring, can be arrived at by applying the same concepts we used when we analyzed the movements from the 50'-radius array to the center of the fire-ring. To produce the final rock arrangement, the distance that each rock must move is half of the 30" diameter, or 15". The probability that any rock could accomplish this would be the reciprocal of the distance, or one chance in 15. The number of rocks that could be placed in our 30"-diameter fire-ring is 10. Hence, the probability that a rock would move in the right direction to take its proper place in the fire-ring would be one chance in ten.

Random Selection of Rocks from the Initial Array

We should now recall that we have 64 rocks in our selection array, which is shown in Figure 9.2, but only 10 of these will be selected for the fire-ring. To accomplish a proper statistical analysis, these 10 rocks must be selected at random from the group of 64. We should not, for example, choose all of them from one circle. For convenience, these rocks have been arranged in a pattern which has located all of them in 6 concentric circles. We must now determine how many rocks should be selected from each of these circles, to insure that our selection process is a random one.

Since there are 64 rocks, and 10 are to be selected, the probability that any particular rock would be selected should be 10/64, or one chance in 6.4. If we multiply this probability by the number of rocks in a particular circle, we can determine how many rocks must be selected from that circle. For example, the circle which has a radius of 400" has 12 rocks in it. Hence, we should assume that the number of rocks to be moved from that circle would be $(12)(1/6.4) = 1.9$. The closest integer to 1.9 is 2. Thus we shall assume that 2 rocks would be moved from this 400"-radius circle to the center of the fire-place. Table 9.4 gives similar data for all of the circles, and its right-hand column specifies the number of rocks to be moved from each of the circles in the selection array.

TABLE 9.4
NUMBER OF ROCKS TO BE SELECTED FROM EACH CIRCLE

A CIRCLE OF ROCKS DEFINED BY ITS RADIUS, INCHES.	B NUMBER OF ROCKS IN CIRCLE	C CALCULATED NUMBER ON ROCKS TO BE TO BE SELECTED FROM THIS CIRCLE.	D ACTUAL WHOLE NUMBER OF ROCKS TO BE SELECTED FROM THIS CIRCLE.
0	1	0.16	0
100	3	0.47	1
200	6	0.94	1
300	9	1.4	1
400	12	1.9	2
500	15	2.3	2
600	18	2.8	3

Summary of All Probabilities Involved in Fire-Ring Construction

We have now established all of the probabilistic ingredients required to calculate the actual statistical probability that natural forces alone could accomplish the design and construction of this fire-ring. According to the theory of probability, if an outcome is dependent upon the accomplishment of several independent events or achievements, the probability that the outcome will be realized is equal to the product of all of the probabilities associated with the several ingredients. Thus, if an outcome is dependent upon the occurrence of two independent events, and the probability of one event occurring is one chance in 3, and the probability of the other occurring is one chance in 4, the probability of the dependent outcome will be $(1/3)(1/4) = 1/12$, or one chance in 12. Table 9.5 below records all of the independent ingredients involved here, together with the probability that each will be accomplished.

TABLE 9.5. SUMMARY OF PROBABILITIES

ACCOMPLISHMENT THAT MUST BE ACHIEVED BY NATURAL FORCES TO COMPLETE THE DESIGN AND CONSTRUCTION OF THE FIRE-PLACE.	PROBABILITY THAT THIS ACCOMPLISHMENT CAN BE ACHIEVED, EXPRESSED, AS THE RECIPROCAL OF THE PROBABILITY.
Proper size rocks (9.4") must be found.	10
Number of rocks to be moved 600".	3
Above rocks must be moved a distance of 600".	600
Above rocks must be moved in the right direction.	401
Number of rocks to be moved 500".	2
Above rocks must be moved a distance of 500".	500
Above rocks must be moved in the right direction.	334
Number of rocks to be moved 400".	2
Above rocks must be moved a distance of 400".	400
Above rocks must be moved in the right direction.	267
Number of rocks to be moved 300".	1
Above rock must be moved a distance of 300".	300
Above rock must be moved in the right direction.	200
Number of rocks to be moved 200".	1
Above rock must be moved a distance of 200".	200
Above rock must be moved in the right direction.	134
Number of rocks to be moved 100".	1
Above rock must be moved a distance of 100".	100
Above rock must be moved in the right direction.	67
Number of rocks to be moved to final place in fire-place ring.	10
Above rocks must be moved a final distance of 15", to ring.	15
Above rocks must be moved in the right direction, to ring.	10

Calculation of Overall Probability

Using the data in the above Table, we can now calculate the overall probability that natural forces could design and construct the fire-ring. The reciprocal of this probability will be the product of the above numbers. This can be calculated as follows, where P will represent the final probability.

$1/P=$
$(10)(3)(600)(401)(2)(500)(334)(2)(400)(267)(1)(300)(200)(1)(200)$
$(134)(1)(100)(67)(10)(15)(10) =$

$$8.32 \times 10^{33}$$

For those of you not familiar with exponents, this is the equivalent of saying that there is one chance in,

$$8,320,000,000,000,000,000,000,000,000,000,000$$

that natural forces could design and create this fire-ring, without the aid of any other designer or craftsman.

Conclusion Concerning the Design and Construction of the Fire-ring

This is essentially the equivalent of concluding that there is no possibility whatever that natural forces alone could have designed and constructed this fire-ring in the wilderness. It had to have been created by intelligent beings, such as human beings. Nevertheless, there could be ants living near the fire-ring who had never seen a human being, and therefore they do not believe that human beings exist. These ants believe that the fire-ring came into existence through natural forces alone. Are you smarter than an ant?

CHAPTER 10. WHO DESIGNED THE PROTOZOA?

Could Natural Forces Make a Fire-ring or a Spider?

At the beginning of this book, I asked you how long it would take you to make a spider out of a rock; and I stated that I didn't think you could ever do it, even if you tried for a million years. Evolutionists believe that a spider can be made out of a rock, possibly with a little air and a little water, by nobody. The process doesn't even need your help. In fact, they believe it has been done, just by the natural forces of the universe. But we showed in the previous Chapter that these natural forces couldn't even make something as simple as a fire-ring. Which do you think is more likely, that the fire-ring was made by somebody, or by nobody?

Could You Make a Protozoan out of Sea Water?

Now I am going to quiz you about a much simpler task than making a spider. How long do you think it would take you to make a single-celled animal, a protozoan, out of sea water. If you need it, you may also use a little air and a few grains of sand. A protozoan is the simplest of all animals. Surely you could make it. It consists of only one cell. Do you think you could make it in a day, a month, or a year. I'll give you all the time you need. Possibly you need to know more about a protozoan before you answer this question. So let's describe this single-celled animal in some detail.

What is a Protozoan?

A protozoan consists of a mass of gelatinous substance called cytoplasm, plus a nucleus within the cytoplasm, and a thin membrane surrounding the cytoplasm and the nucleus. But cytoplasm is not just a homogeneous inert substance. Recent research has shown that within what scientists have in the past called cytoplasm, there are many complex structures and much activity. Many protozoa have hair-like appendages protruding from the outer membrane which act as oars with which the animal can propel itself. Most protozoa live in fresh water, the oceans, moist soil, or inside other animals. They seem to need a moist environment.

The typical size of a protozoan is about 150 micrometers long. A micrometer is one millionth of a meter, and a meter is about 39 inches. Therefor, a typical protozoan would have a length, L, of about,

$$L = (150/1,000,000)(39) = 0.0060 \text{ inch.}$$

This is six thousandths of an inch. I measured the diameters of several human hairs. They averaged about 0.0015 inches in diameter. So the protozoan is about as long as the thickness of four hairs. This is a microscopic dimension. It is very small.

The Protozoa are Extremely Complex

Although the protozoa are the simplest of all animals, they are, actually, extremely complex. In multicellular animals there are organs and parts each of which has a specific function to perform. The organs and parts employ specialization. For example, legs provide locomotion, the digestive system digests food, the nervous system provides a sense of touch, the respiratory system takes in oxygen-laden air and expels waste gasses. The reproductive system produces offspring. Each of these systems consists of may cells, each of which specializes in joining with other cells to produce a specific function.

Believe it or not, in a protozoan, all of these functions are performed by the protoplasm of which the cell is composed. Different regions within the cell specialize and perform these functions which are necessary for life and continued existence. These specific regions are called organelles. Particular organelles may perform the following functions: provide sensory capability, serve as a contractile system, provide for locomotion, provide self defense, eat food, digest food, excrete wastes, bring in oxygen, expel carbon dioxide, provide for reproduction, etc. More complicated animals have specialized parts and organs to provide these functions. But protozoa can perform the same functions using specific regions of the protoplasm within a single cell. As you estimate the time it would take you to make a protozoan, figure out how you could make a single cell which can perform all of these functions. It is, indeed, a formidable task. A single-celled animal is not a simple creature.

Locomotion by Flagella and Cilia

Now let's consider in detail just one of the above functions, locomotion. Many protozoa have oars which they can move to propel themselves in water. Some of these are long and whip-like, called flagella; others, called cilia, are shorter, and are usually more numerous. Flagella may be about 150 micrometers long. Cilia may be about 5 to 10 micrometers in length. This is about 0.0003 inches.

If one were in a row-boat and he took an oar and moved it back and forth behind the boat, the boat would move forward. A sailboat can be caused to move forward merely by moving the rudder back and forth. If a person who is a scuba diver were to put rubber flaps on his feet and move them back and forth, he would be propelled forward. Similarly, flagella at the back end of a protozoan can move with a back-and-forth whip-like motion, and this will propel the protozoan forward.

Alternatively, if a person were in a rowboat and he rowed the boat, having the oars oriented outward perpendicular to the boat, he could move the boat forward. The rowing action of cilia located at the sides of a protozoan can move the animal forward. Both flagella and cilia are found in multicelled animals throughout the animal kingdom; and in almost all cases, they are of the same identical design. In fact, the movement of a human sperm cell, seeking to find an egg, is by the action of flagella on the back of the sperm cell.

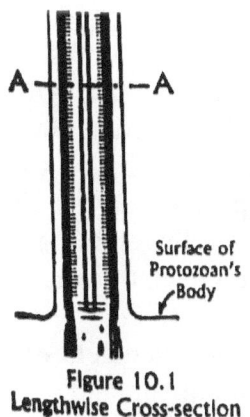

Figure 10.1
Lengthwise Cross-section
Through a Flagellum

Microtubules Move the Flagella and Cilia

Let's now consider in more detail just how these flagella and cilia work. Figure 10.1 shows a split-hair-type cross-section through a flagellum. This flagellum is a hair-like appendage pro-truding outward from the surface membrane of a protozoan. Figure 10.2 shows a cross-section through plane A-A of the flagellum. I hope you recognize the similarity of this cross-section to our fire-ring on the mountain-top, discussed in Chapter 9. Inside the flagellum are 10 pairs of microtubules. These are arranged in a circle of 9 pairs of microtubules,

99

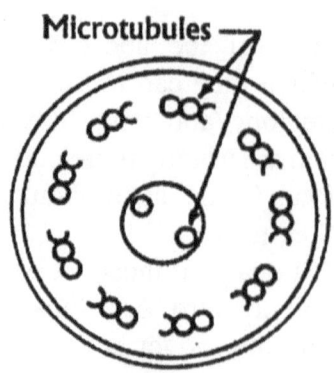

**Figure 10.2
Cross-section Through
Plane A-A of Fig. 10.1**

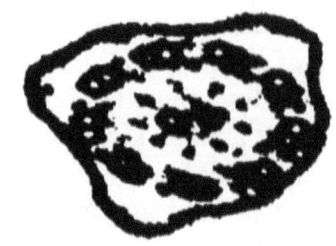

**Figure 10.3
Actual Photograph
of a Cross-section
Through a Flagellum**

plus a tenth pair at the center. Our fire-ring had 10 rocks arranged in a circle.

It should be of interest to us to learn the size of these hairs called flagella. They are very small. Figure 10.3 is an actual photograph of a cross-section similar to that of A-A in Figure 10.2, taken with an electron microscope, at a magnification power of 133,000, on the original Photograph. I measured the diameter of this photo-graph of a flagellum at six different locations and found that its average diameter, on the picture, was 23 millimeters. Based on these measurements, we can calculate the average diameter of the flagellum to be,

$$D = (23/133,000)(1/25.4) = 0.000,006,8 \text{ inches.}$$

This is about 7 millionths of an inch. Bear in mind the extreme minuteness of this flagellum as you estimate how long it would take you to make one protozoan. If the length of a protozoan were about 0.006 inches, it can be seen that the diameter of one flagellum is only about one thousandth of this length. It is, indeed, a mere hair on the body of the single-celled protozoan.

Now, if we scale, from Figure 10.2, the relative diameters of the microtubules as compared with the flagellum, we can conclude that a microtubule is about 1/19th of the size of the flagellum. This means that the diameter of one microtubule would be,

$$d = (1/19)(0.000,006,8) = 0.000,000,36 \text{ inches.}$$

This is about one third of a millionth of an inch. You will need very small fingers to make these tubes. Next, let's consider how these flagella actually work Suppose you were to cut two lengths of garden hose, each

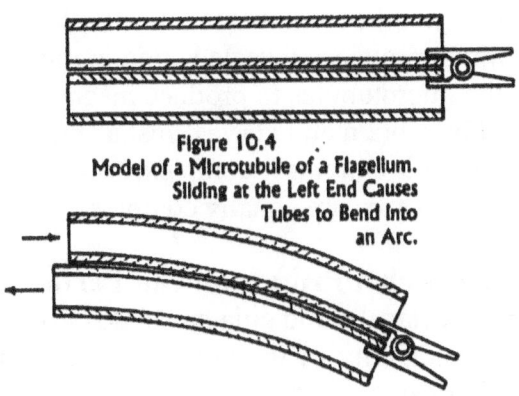

Figure 10.4
Model of a Microtubule of a Flagellum.
Sliding at the Left End Causes
Tubes to Bend Into
an Arc.

length to be about 20 inches long. Now align the two pieces of hose side-by-side, as shown in Figure 10.4, and fasten the two hoses together at one end, using a spring-loaded clothes pin. We now have a model of a pair of microtubules, similar to those shown in Figure 10.2. Now take the other, non-clamped, ends of the two hoses and slide one hose longitudinally a small distance with respect to the other. Notice that the clamped ends of the pair of tubes moves laterally. This is how the pairs of microtubules cause the flagellum to move. By coordinating the sliding movements of all nine pairs of microtubules, the flagellum can be made to move in any particular desired direction, and with particular desired timing of the movements. In real life these movements and the timing are coordinated to produce whip-like motions of the flagella which then propel the animal in water.

Muscles and Nerves Power and Control the Microtubules

It remains for us to explain how the sliding movements at the bases of the microtubules are made to occur. They are accomplished by the same mechanism that causes the muscles of all animals to move and pull on tendons. Contrary to a popular conception, muscles do not contract simply as a result of the radial expansion of the muscle fibers. Recent evidence indicates that muscle fibers actually slide one with respect to another to produce a shortening of the muscle, which produces a pull on the tendon. This sliding action involves a complex integration of electrical impulses and chemical reactions, the overall result of which is to transform the chemical energy in blood sugar and oxygen into the mechanical work of the sliding action. The sliding of the microtubules, one with respect to another, is, in fact, a muscle-like phenomenon identical to that which produces muscle action in multi-celled animals; and, in order for the microtubules to produce meaningful motions, they must be controlled by a brain and nerves. Just as the science of muscle action in larger animals is extremely complex, so also the movements of flagella and cilia in protozoa are equally complex.

Other Functions which a Protozoan can Perform

This discussion of flagella and cilia has to do only with the function of locomotion, the capability of the animal to move. There are other functions also which can be performed by this animal, including: the acquisition of food, the digestion of food, the acquisition of oxygen, the excretion of wastes, self defense, sexual activity, and reproduction. Some protozoa even have a region which serves as an eye with which they can detect and respond to light. We will not take the time here to discuss these other capabilities of the protozoa, but please be assured that each of these systems may be just as complex as the above-described system for locomotion.

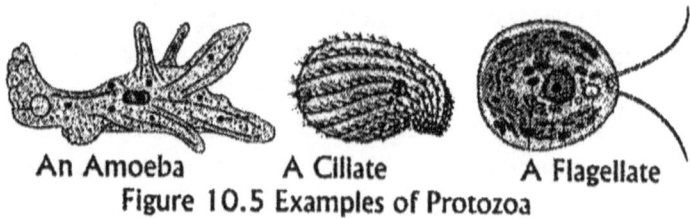

An Amoeba   A Ciliate   A Flagellate

Figure 10.5 Examples of Protozoa

Protozoa are Extremely complex

There are more than 65,000 different known species of protozoa. Some of them have already become extinct. Obviously, if I were to ask you to design and construct each of them, it would take you thousands of times as long as it would take you to make one of them. Figure 10.5

102

shows several protozoa. Do they look like simple elementary blobs of chemical substances, not much different from the inorganic contents of early-earth sea water? If you think that, then please quickly answer the question posed at the beginning of this Chapter. Protozoa are not simple. They are extremely complex! You couldn't make one in a million years.

Were Protozoa Produced by Lightning Striking Sea Water?

The primary objective of this book is to speculate on the origin of animals. Who designed them? Who constructed them? Evolutionists believe the following scenario. The atmosphere of the primitive earth, before animals appeared, was believed by some scientists to consist largely of carbon dioxide, hydrogen, nitrogen and water, with possibly some methane and some ammonia. The oxygen we have today was believed to be missing then. When lightning struck these substances it is believed that some chemical compounds related to life might have been synthesized. Experiments using electric sparks were found to produce such chemicals, including amino acids, adenine (a component of ATP), nucleic acid bases, and sugars. Then, it is speculated, such chemicals were concentrated by various means, in special places of the earth's environment, to produce larger molecules that ultimately resulted in proteins and nucleic acids. Having produced, on the primitive earth, some of the chemicals that are similar to some of the molecules found in living animals, it is then speculated that, somehow, these chemicals became transformed into a protozoan, the first, and simplest, of all animals, possibly by the action of lightning striking the concentrated sea water. Then, of course, the program of mutations and natural selection came into play and all of the other more complicated animals evolved.

Could Natural Forces Design and Construct a Protozoan?

Let's now estimate the statistical probability that the protozoa which we have studied in this Chapter could have been designed and constructed by the natural forces of the universe, with no help from any intelligent being. First, let me again call to your attention the striking similarity between the circular design of the microtubules shown in Figure 10.2, and the rocks of the fire-ring shown in Figure 9.1. In Chapter 9 we carefully analyzed this fire-ring and came to the conclusion that the statistical probability that it could have been designed and constructed by the natural forces of the earth, alone, was one chance in,

8,320,000,000,000,000,000,000,000,000,000,000.

By what process of reasoning could you conclude that the ring of microtubules of one flagellum of a protozoan would have any better chance of being made by natural forces alone, than would the ring of rocks in the fire-ring? The rocks of the fire-ring came from a circular region having a radius of 50 feet. Where did the atoms and molecules come from which make up the microtubules of a flagellum? We assumed that each of the rocks was a completed entity, and all we had to do was move it a few feet in the right direction. A protozoan is made up of about 62% oxygen, 17% carbon, 12% hydrogen, and lesser percentages of nitrogen, sulfur and other elements. The atoms of which each microtubule is made had to first participate in a series of complex chemical reactions before it could become a part of one of the molecules of a microtubule. And where was the place of origin of each of these atoms, just before it started to enter into the process by which it ultimately became a part of the protozoan? If the protozoan lived in sea water, these atoms came from the sea, or the air above it. Some atoms may have come from land minerals which were washed into the sea.

Let's now visualize a protozoan in the ocean. Let's assume it is the first one that ever existed. Before it became a protozoan, each of its atoms was somewhere else in the ocean. These atoms had to come together, combine chemically to form molecules, and then become part of the living protozoan. To estimate the statistical probability that these atoms could become a protozoan without any assistance from any intelligent designer-craftsman, I as an engineer, am going to make several assumptions, and then perform some simple calculations. Basically, I am going to calculate how many atoms there might be in a protozoan, and then I am going to assume that the probability that any one of these atoms would move from its location in the sea to become a part of the protozoan, is about the same probability that a rock of Chapter 9 would move to, and become a part of, the fire-ring. Actually, I think this is a very conservative assumption, because the sea is very large, and the protozoan is very much more complicated than the fire-ring. I will assume that the number of molecules in a given volume of the protozoan is about the same as the number of molecules in an equal volume of water. This should be a reasonable assumption, because much of a protozoan is water. Most animals are largely water.

The molecular weight of a molecule is the sum of the atomic weights of its component atoms. The atomic weight of hydrogen is about 1. The atomic weight of oxygen is approximately 16. Since a molecule of water contains one oxygen atom and two hydrogen atoms, the molecular weight of water is about 18. Consider a quantity of water the mass of which is the same number of grams as its molecular weight, that is 18 grams. This is called a mole of the substance. The number of molecules in a mole of a substance is the same for all substances. This number is called Avogadro's Number, and it is

$$6.02 \times 10^{23}.$$

Thus, the number of water molecules in 18 grams of water would be, 6.02×10^{23}. Water weighs one gram per cubic centimeter. Hence the number of molecules of water in one cubic centimeter would be,

$$N = (6.02/18)(10^{23}) = 3.34 \times 10^{22}.$$

If we recall that our protozoan is 150 micrometers long, and if we assume that our protozoan is approximately of a cylindrical shape, and is half as wide as it is long, its volume could be calculated as the product of its length times its cross-sectional area, or,

$$V = LA.$$

The length of the protozoan, in centimeters, would be,

$$L = (150)(100)/1{,}000{,}000 = 0.015 \text{ centimeters.}$$

Its diameter would be half its length, or, $D = 0.0075$ centimeters. Its cross-sectional area, A, would be,

$$A = (3.1416/4)(D^2) = (3.1416/4)(0.0075)^2$$

$$A = 0.000{,}0442 \text{ square centimeters.}$$

Then,
V = (0.015)(0.000,0442) = 0.000,000,663 cubic centimeters

The number of molecules in our protozoan would then be,

$$M = (3.34 \times 10^{22})(0.000,000,663) = 2.21 \times 10^{16} \text{ molecules.}$$

Since there are three atoms in each water molecule, then, based on water, the number of atoms in each protozoan would be,

$$A = 6.63 \times 10^{16} \text{ atoms in one protozoan.}$$

Finally, since we have in our protozoan, 6.63×10^{16} atoms, and in our fire-ring we had 10 rocks, then the statistical probability that a protozoan could have come into existence by natural forces, alone, would be given by,

$$1/P = (8.32 \times 10^{33})(6.63 \times 10^{16})(1/10) = 5.52 \times 10^{49}.$$

This means that there would be one chance in,

55,200,000,000,000,000,000,000,000,000,000,000,000,000,000,000,000

that a protozoan could have been designed and constructed by the natural forces of the universe, with no help from any intelligent designer-craftsman. This is the equivalent of saying that it didn't happen. Other calculations that take into consideration the vast extent of the oceans, and the billions of years that the earth has been in existence, still show that there has been essentially zero probability that even one protozoan could have arisen by chance.

Natural forces, alone, did not create the protozoa. The forces alleged to be possessed by evolution could not possibly have brought into existence a protozoan. And this is the simplest of all animals, and is thought by evolutionists to be the father of all of the other animals. Recall now that evolutionists believe that, since some of the molecules found in living matter might have been synthesized by lightning striking the early earth, therefore, they think, it is logical to believe that, somehow, a creature as complex as a protozoan could suddenly, or even slowly, arise from sea water. Our analysis here shows that this belief is incredible.

My Opinions as an Expert Witness

Let's remember now that you are the Judge and Jury, and I am the expert witness. I have just presented to the court my engineering analyses and my findings relative to the protozoa. I must now be asked this question,

> "In your opinion, as an expert witness, and based on a reasonable degree of engineering certainty, who do you think designed and constructed the protozoa? Were they designed and built by some intelligent designer-craftsman, or were they designed and built by nobody, just the natural forces of the universe?"

My answer to this question is that the protozoa were clearly designed and constructed by some intelligent designer-craftsman!

"And what is the basis for your opinion?", I should be asked. To answer that question I will need to summarize this Chapter.

First, I have shown that there is only one chance in 5.52×10^{49} that they could have come into existence by natural forces alone. Secondly, as an engineer, when I see a complex pattern which involves non-random arrangements of materials, and which has an obvious purpose and function to perform, I see the work of an intelligent designer. In the protozoan, there are hair-like appendages, flagella, which have the obvious purpose of enabling the animal to move about in water. Each of these flagella has within its tubular housing, 10 tiny tubular-shaped pairs of microtubules. These are like muscles and can cause controlled movements of the flagella. The exterior tube of the flagella is basically circular, not of random shape. Each of the tiny microtubules is round in shape, and each is only 360 billionths of an inch in diameter. They are arranged in a neat circle, just like the rocks of our fire-ring. Clearly, I see a non-random pattern; and the entire flagella is an ingenious mechanical device and it has a mechanical purpose. It was clearly designed by some intelligent being. It could not have evolved. And, the other parts of the protozoa are equally remarkable and equally deserving of the conclusion that they were designed by some very skilled designer-craftsman. These are my opinions, as an expert witness.

What is Your Opinion?

Now that you know more about a protozoan, how long do you think it would take you to make one out of some sea water? How long do you think it would take for this protozoa to be made by nobody? Don't you think that you would have a better chance of making a protozoan than would nobody? Do you think the protozoa were designed and constructed by nobody?

Was my 1926 Truck Produced by Lightning Striking the Mesabi Range?

In my garage I have a 1926 Model TT Ford truck. It is made almost entirely of iron. In northern Minnesota, in the Mesabi Range, in the twenties, there was a lot of iron ore, iron oxide. Do you think that a few bolts of lightning hitting the Mesabi Range might have transformed some of that iron ore into a Model TT Ford truck? I think there is just as much chance of that happening as there is of lightning acting on some concentrated sea water and producing a protozoan, many years ago. I can study my Model TT Ford truck, and see abundant evidence of the work of a designer and a skilled craftsman. I see the same evidence as I study the protozoa.

1926 Model TT Ford One-ton Truck

Why Turtles Believe in Evolution

When I was a boy, my brother and I enjoyed observing all of the different kinds of automobiles which we saw. In fact, we kept a written record of all of these automobiles. We saw Fords, Buicks, Dodges, Kissels, Durants, Essex, Hupmobiles, LaSalles, Marmons, Overlands, Whippets, and many others. Similarly, recently, on a rock a few feet below the surface of the ocean, two young turtles entertained themselves by observing the spectacular number of different-looking protozoa that swam by. One said to the other, "I have seen 10,325 different protozoa, and they are still swimming by." The other turtle said, "I wonder where they come from. They seem to go by in an endless variety." After further discussion, they both agreed that neither of them had ever seen anybody that was designing and constructing protozoa. So they concluded that *no designer of protozoa exists!* Based on that assumption, they were forced to conclude that the protozoa came into existence by the actions of the natural forces of the earth, by evolution. They could come to no other conclusion, because of their assumption that no designer existed.

Are You Smarter than a Turtle?

Do you now know how long it would take you to make a protozoan out of sea water? What do you think? Are protozoa complex? Were they designed and constructed by somebody? Or were they designed and constructed by nobody? Which is the more logical conclusion? Are you smarter than a turtle?

The Cambrian Explosion

CHAPTER 11. SPIDERS AND WATCHES

Fingerprints Found on Spiders and Watches

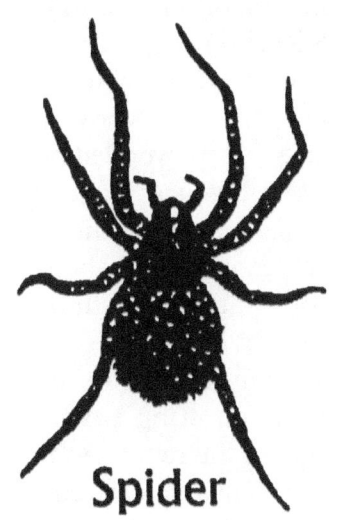

Spider

Have you by now figured out the answer to the question posed in Chapter I of this book, "How long would it take you to make a spider out of a rock?" Hopefully, after you read this Chapter you will be better able to answer this question. But, first, let me remind you that, under the format of this book, I am the expert who will be analyzing machinery, including animals, and I will give expert testimony concerning who designed and built these machines and animals. In Chapter 5 we identified the primary clues that I look for to determine how the animals were created. These are the fingerprints of the designer. We will now apply the contents of Chapter 5 as we consider various animals, or parts of animals, to determine their origins. We have already studied the smallest animal, a protozoan, as well as an inanimate structure, a fire-ring on a mountain top. We should now move on to considerations of larger animals. Let's next consider an animal much larger than a protozoan, a spider. Our fingerprints of Chapter 5 apply to man-made machines as well as animals, so let's add to our study a small machine, the wristwatch, in addition to the small animal, the spider.

Watchmaker Throws Watches into Ravine

This Chapter will contain mostly facts, but, to introduce some new concepts, it will also contain some fiction. I assume you will be able to distinguish between the facts and the fictional content.

Raccoon

On the eastern edge of a small midwestern city there was the shop of an elderly watchmaker. A watchmaker is one who makes, or repairs, watches. His shop was so located that the back window of his building overlooked a large, deep, wooded ravine.

Over the years that the watchmaker had been in business there, from time to time, watches had been brought to him for repair which were so badly broken or worn-out that they could not economically be fixed. In many such cases, sometimes with disgust, our watchmaker threw the watches out of his back window into the wooded ravine. Thus, scattered about in the floor of the woods below our watchmaker's shop, were a wide assortment of watches, with representatives covering a period of many decades in the evolution of watches. Also in the ravine, at night, as in most such natural habitats, there were raccoons, which wandered about in search of food.

Zoology Professor Studies Spiders

In the same City there was a large University, and in that University there was a Department of Zoology. One of the professors of that Department specialized in Arthropods; in particular, arachnids, especially spiders. Since there were many spiders in the wooded ravine behind the watchmaker's shop, this professor, and his graduate students, were frequent visitors to the ravine. There they studied the spiders. So now we have a wooded ravine filled with spiders and watches. Before we continue with our story, however, I think we should study each of these small machines in some detail.

The Evolution of Watches

Breguet (1815)

Let's study watches first. Since watches are machines, and since my field of expertise is machine design, I know something about watches. Also, I have a friend who is a watchmaker and I have access to his library. Scattered throughout this Chapter I have inserted pictures of some of the watches which were thrown into the ravine by the small-city watchmaker of our story.

Spring-driven timepieces first appeared in Europe in about 1430. By 1845 watches were being made in America. Most watches designed and constructed between 1430 and 1957 consisted of a delicate balance-wheel, a spirally-wound hair-spring, an escapement mechanism, a main spring, and many shafts and gears. The timekeeping capability of these watches was based on the oscillations of the spring-mass system, the balance-wheel and the hair-spring, vibrating at its natural frequency and controlling the motions of the escapement mechanism. The source of the energy in these watches was the main-spring, which had to be wound each day. More recently, some of the watches have had spring-mass systems made of tuning forks, or electrical vibrations in quartz crystals or other circuits. Many of the parts of these watches can be seen in the pictures of this Chapter.

Trenton (1895)

Throughout the period between 1430 and 1957 many improvements were invented and applied to the various watches, such as the use of jewels (1700), new types of escapements (1759 and 1799), stem winding (1824), the invention of Invar (1913), shock-resistance (1930), and the unbreakable main-spring (1948). In 1957 Hamilton developed a wristwatch that was powered by an electrical battery rather than a main-spring. In 1960 Bulova introduced the Accutron, using transistors and a tuning fork. In 1968 quartz analog watches were developed. In 1970 solid-state digital watches were introduced. A watch was developed in 1985 which generated electricity by the motion of the user's arm, which was then stored in an electrical condenser. So you can see that these watches evolved substantially over the years. And, as the raccoons came out at night in the ravine, they studied these watches and noted their evolutionary changes.

The Remarkable Anatomy of a Spider

Elgin (1900)

During the daytime, and often at night, the professor and his graduate students came to the ravine and studied the spiders. So let us, also, at this time, study the spider. Figure 11.1 shows the cross-section of a spider. (Hickman 2) Generally, the body of a spider consists of two parts, the front part, called the cephalothorax, and the rear part, called the abdomen. They are fastened together by a narrow waist, called the pedicel. Spiders range in size from a length less than one millimeter (0.040 inches) to about 3.5 inches. Spiders have eight legs, plus four other appendages, all of which are attached to the cephalothorax. They have an external skeleton with the soft parts of the body housed within the outer shell structure. The legs are segmented hollow tubes with muscles within. Believe it or not, spiders have a heart, a brain, eight eyes, lungs, sex organs, a mouth, a digestive system, an anus, and an extensive network of nerves, blood vessels, and muscles. These are all shown in Figure 11.1.

In addition, spiders have several organs which are special; they have silk-producing spinnerets, and poison-injecting fangs. At the posterior end of the spider's abdomen there are six spinnerets, each of which is connected by hundreds of microscopic tubes to silk-producing glands in the abdomen. At a signal from the spider's brain, muscles squeeze a liquid secretion, a protein called fibroin, through the tubes toward the spinnerets. The flow rate is controlled by muscular valves in the tubes. As the liquid secretion exits from the spinnerets it solidifies and becomes a

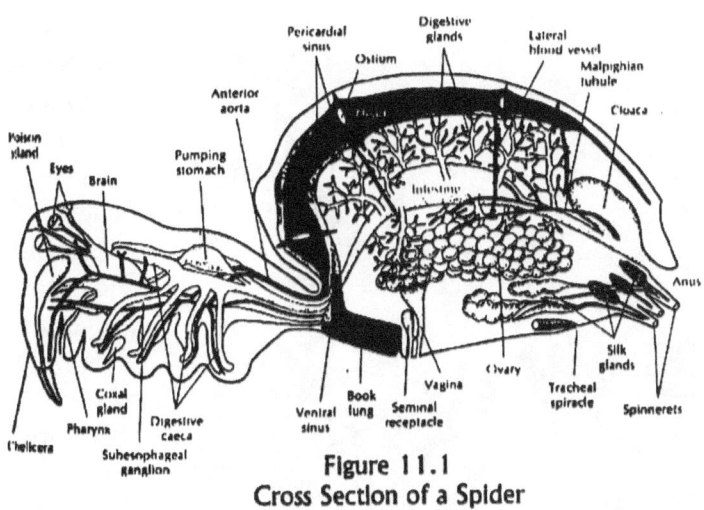

Figure 11.1
Cross Section of a Spider

thread-like filament. The spider can control the spinnerets to produce a single large filament or several smaller ones. The strength of the silk thus formed is remarkable, being actually stronger, in terms of strength per unit weight, than steel. The silk extruded from the spinnerets can be used to form beautiful geometric webs, in which prey, usually insects, can be trapped. The silk can also be used to tie up trapped prey, to construct egg sacs or sperm webs, or to "balloon", ie. to fly through the air, blown by the wind on a long strand of silk. The spider can spin silk filaments which are not sticky or others that are sticky. To build a geometric web the spider selects a spot, such as the site in a tree shown in Figure 11.2, and he then climbs to some high spot and extrudes out a filament which the wind blows horizontally until it sticks to some other object a foot or two away, such as another branch, as shown in Figure 11.3. This is the first strand of the web. The spider then walks back and forth on this strand, extruding other strands, to reinforce this initial primary filament. The spider then goes to the middle of the strand and descends downward on a second filament, until he hits the ground or another branch, as shown in Figure 11.3 This is the second strand.. By similar gymnastics the spider then installs other radially oriented strands, until the web looks like Figure 11.4. All of the strands installed thus far are of the non-sticky variety. For the final phase of web building, the spider goes to the middle of the structure and lays a sticky filament in a spiral formation as shown in Figure 11.5.

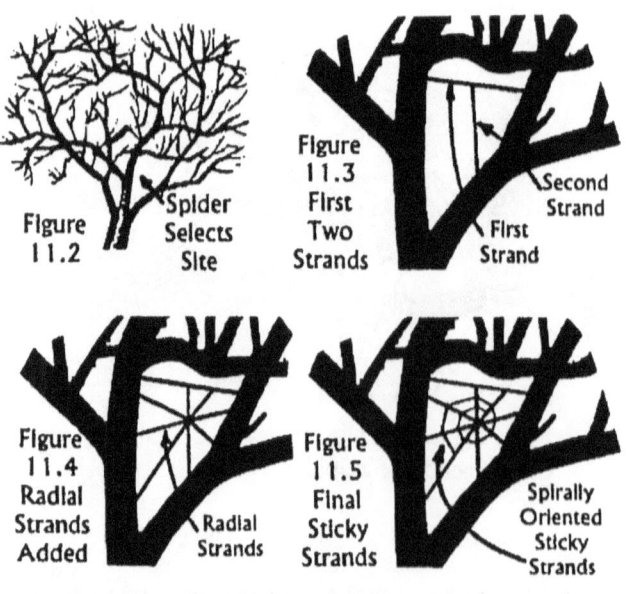

Figure 11.2 Spider Selects Site

Figure 11.3 First Two Strands

Second Strand

First Strand

Figure 11.4 Radial Strands Added

Radial Strands

Figure 11.5 Final Sticky Strands

Spirally Oriented Sticky Strands

The other remarkable device possessed by the spider is the pair of poison fangs located at its front end. Inside of the spider's head, below its eyes and in front of its brain are two poison glands. These are shown in Figure 11.1. These glands produce a poisonous venom which is usually a nerve toxin. Each of these glands is connected by a duct to a sharp, pointed, needle-like appendage which the spider can use to pierce the prey or enemy. When the hollow fang pierces the enemy the spider can force some of its venom into the body of the victim. The venom can kill or disable an insect, and, in the case of some spiders, such as the black widow, it has been known to kill human beings.

Fingerprints of a Designer Found on Spinnerets

Howard (1915)

Now let's apply some of the concepts of Chapter 5, on fingerprints, to analyze the spider. Let's consider, first, the silk-producing spinneret devices. The overall system consists of the silk glands, the microscopic tubes, the array of six spinnerets, and the muscles and nerves that control the system. All of these parts obviously constitute a very clever, complex device that would clearly be patentable. It is novel, ingenious, and useful. This spinneret system is clearly a multipart system. It would be worthless if any single part were missing. All the parts must work together to achieve the objective of making various types of silk. The flow of the liquid through the tubes, being controlled by muscular valves, and the six spinnerets, clearly make this a complex mechanical system of the fluid-flow type. The chemistry involved is even more remarkable. This creature eats insects, and out of this raw material it is able to apply some extremely complicated chemical reactions to produce a protein liquid which, when extruded out of its spinnerets, produces a silk thread that is stronger than steel. The parts of this extrusion system, especially the tiny hollow tubes, clearly have shapes that suggest purpose and design. And, finally, when the spider spins a beautiful web having radial lines and spiral netting, an artistic pattern has clearly been created. The six spinnerets obviously constitute an ordered array, not a random pattern. The nerves which control all of this are, of course, electrical devices of ingenious design. Thus, all of the fingerprints of a designer are spectacularly present in the spider's silk-spinning facility.

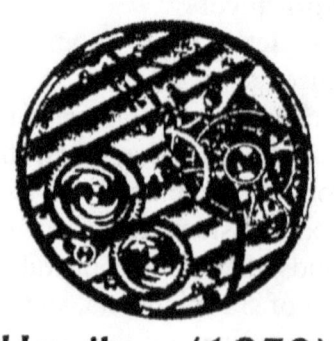

Hamilton (1930)

You may now ask me the question. I am the expert witness. You are the Judge and Jury. Do I see the fingerprints of a designer in the spinneret system of a spider? You bet I do! Is it my testimony that, to a high degree of engineering certainty, these spinneret systems had to have been designed and constructed by some extremely competent designer? The answer is clearly, "Yes!" If you doubt this, I suggest that you go back and restudy the fingerprints of Chapter 5, and see if you agree that they are all clearly present, and sensationally applied, in these spinnerets.

Fingerprints of a Designer Found on Poison Fangs

Now let's analyze, in terms of fingerprints, the poison fangs on the front of the spider. Here, again, each of these fangs is clearly a very clever, patentable device. A substance which will affect the nervous system of the spider's prey is chemically synthesized and delivered to the prey through a tubular duct in a sharp fang built like a hypodermic needle. This is, indeed, a very ingenious and effective device. The spider eats insects, and from this raw material the poison glands transform this substance, using complex chemical reactions, into a poisonous venom that will kill the spider's prey. It is a mechanical system involving flow through tubes, and the squeezing action of specially designed muscles, controlled by electrical nerve impulses. Each poison fang is obviously a multipart system, and it uses shapes that are clearly designed for the purpose of the application, tubes and sharp pointed needles. Most of the fingerprints identified in Chapter 5 can be found in these fang systems. The spider's devices for creating and injecting venom are also truly spectacular contrivances.

**Pierce (1955)
Winds Automatically**

Again, in this case, my expert-witness testimony is that this poison-fang system could not possibly have been conceived, designed and constructed solely by the natural forces of the earth. Based on my analyses, involving the fingerprints outlined in Chapter 5, the spider's venom-injecting facility clearly had to be designed and constructed by some intelligent designer-craftsman.

How Long Would it take You to Make a Spider?

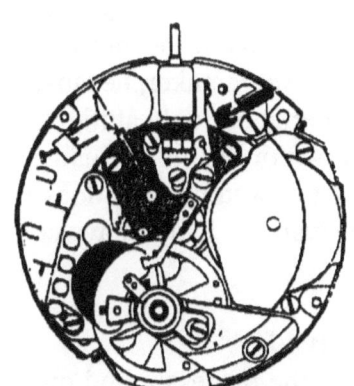

**Bulova (1965)
Uses Transistors
and Tuning-fork**

Now that you know more about spiders, can you tell me how long it would take you to make a spider out of a rock? Look, again, at Figure 11.1. Note all of the components and systems that the spider possesses. Could you take a rock, with a little air and water, and produce with them a heart, a cardiovascular system, lungs, sex organs, a nervous system, an exoskeleton, a system of muscles, eight eyes, and a brain, in addition to a silk-producing system and a couple of poisonous fangs, all in a creature that is one millimeter (0.040 inches) in length? How long would it take you to do this? I suspect that you know the answer to this question. You could *never* do it.

Mother Raccoon Believes that the Watches Evolved

Now let's return to our watchmaker's back yard, the wooded ravine. Night after night a mother raccoon and her only remaining baby wandered about in the ravine, seeking their food. Of course, they often observed the watches that the watchmaker had thrown out of his window. One night the baby raccoon asked his mother, "Where do these watches come from, mother? Who made these watches?" The well educated mother raccoon replied, "They evolved, by natural selection, from the ground. The elements of the earth somehow came together to produce these watches." She then went on to explain, "It took many millions of years, but they adapted to this environment. They evolved from the rocks of the earth, by natural selection." The young raccoon responded, "But mother, I have examined these watches. They are very complicated. I don't see how they could have evolved. How do you know for sure that they evolved from the earth?" The mother replied, "I know that they evolved because I have never seen anyone making them, and I have never seen anyone who would be smart enough to make them. Since I have never seen a watchmaker, I am convinced that watchmakers do not exist!" Unfortunately, the raccoons had never seen the watchmaker, because he works there during the day, and the raccoons come out only at night. So the baby raccoon had to agree that, if neither he nor his mother had ever seen a watchmaker, then it must follow that the watchmaker does not exist. And if he does not exist, obviously, the watches had to have come into existence by the process of evolution.

Professor Believes that the Spiders Evolved

Optel (1972)
Digital Quartz with
Liquid Crystal Display

The next day the professor and one of his graduate students came to the ravine to study the spiders. The graduate student asked the professor to explain to him the origin of the spiders. The professor then expounded to his student, saying that the spiders evolved from the elements of the earth. He explained that it took millions of years, and it was done by natural selection. They adapted to the environment. The student, observing how complicated the spiders were, was not satisfied. So he pressed his professor further, asking, "How do you know with such certainty and assurance that the spiders were not created by some very smart designer?" The professor explained in more detail, asserting emphatically that he had never seen any such designer, and he was therefore convinced that no such designer existed. The professor appeared to be annoyed by the audacity of his student to even suggest that the spiders could have been created by some supernatural designer. He repeated to his student, speaking with great emphasis, saying, "I have never seen any supernatural being, and I have never seen any designer of spiders, and therefore I am convinced that no such person exists. If no supernatural designer exists, then I have to conclude that the spiders were designed and created by evolution. A person that does not exist cannot design a spider!"

Are You Smarter than a Raccoon?

Rolex (1980) Chronograph Winds Auromatically

Now, my dear reader, who do you think designed the spiders? Who do you think designed the watches? Could the watches have come into existence by the actions of the natural forces of the earth acting on the mud in the ravine? Are spiders less complicated than the watches? As an engineer who has studied both of them, I can assure you that spiders are very much more complicated than watches, and both of them were designed and constructed by some intelligent being! Are you smarter than a raccoon?

The Cambrian Explosion

CHAPTER 12. BIRDS, BEES AND BATS

Animals That Can Fly

This Chapter will be devoted to animals that can fly, such as birds, bees and bats. These animals have been so designed that they can actually defy the force of gravity, and, by applying the laws of aerodynamics and fluid mechanics, they can rise above the surface of the earth and fly through the air.

How Birds Fly

Let's first analyze the flight of birds. When I was five years old our family moved to a new house, and, as my father drove our car into the new garage, he effectively trapped a chicken in the garage. As the car moved forward, the chicken, apparently thinking that she would be crushed by the car, flew vertically upward about six feet, and she landed on the hood of the car. I was impressed by this performance, and since then I have observed many birds fly directly upward; or, when landing, I have seen them fly slowly downward. And these feats seem to be accomplished by the flapping of their wings up and down. As I write this, I am sitting in my boat, docked at Pelee Island in Lake Erie, and I can observe sea gulls taking off from, and landing on, the large rocks at the edge of the shore. They clearly can go straight up, or land by going straight down.

To the casual observer who has seen birds fly, it would seem that, by the flapping of the wings up and down, a lifting force is developed that opposes the force of gravity. Indeed, I have seen motion pictures of men who have fastened sheets of plywood to their arms and have tried to fly by flapping the plywood sheets. I have seen them jump off of roofs, so equipped, in an effort to fly. All such experiments have ended in total failure, if not disaster.

To the more astute observer, it should be obvious that the flapping of simple wings up and down will not produce any net lifting force at all. When the wings are moved downward, a lifting force is obviously produced; but when the wings are moved upward, a downward force of equal magnitude is developed. These two forces cancel each other, and no net upward force will be produced. But this is exactly what birds appear to do. They flap their wings up and down, and they fly.

This engineering enigma has puzzled me for many years, and so, as a part of my research related to the writing of this book, I made a determined effort to find out, really, how birds do defy gravity and fly through the air. I consulted aeronautical engineers and biologists. I bought books on how birds fly, and on insects in flight. (Terres 1) (Brackenbury 1) (Cromer 1) I bought the tape of a PBS program on how birds fly. (Alda 1) I've had biologists search the subject for me using computers. And I have devoted a lot of time to my own engineering analyses of this subject, but I could not find anyone who could really explain to me how birds fly.

Most of these sources point out that the wings of birds are like the wings of an airplane. They discuss the classic airfoil cross-section, and assert that birds' wings are good airfoils. They suggest that the outer portions of a bird's wings act like propellers, and the inner portions act like the wings of an airplane. The outer portions do not rotate like the propeller of an airplane; they move up and down. They are reciprocating propellers. But they do propel the bird forward. The inner portions then act like the wings of an airplane to generate lift. And I agree with these concepts, but only for high speed flight. They do not explain how a bird can fly directly upward or downward by flapping its wings. One of the books that I read states that "wings can only generate lift once a flow of air has been established across their surfaces". Most of the other sources make similar statements. So, these concepts explain how birds can generate lift under certain circumstances, but they do not explain how a bird can develop lift by flapping its wings up and down.

In a typical flight, a bird flaps its wings rapidly and it rises, often at a steep angle upward. It can even rise directly upward. Then, it gains forward velocity and flies horizontally for a while. It might then glide horizontally with no wing flapping. But then, if it wants to gain more altitude, wing flapping may resume and the bird may soar to a higher elevation. As it glides downward for the purpose of landing, it ceases the wing flapping until it is close to the selected landing site; it glides. Then, just prior to landing, it again flaps its wings to produce enough lift to accomplish the final slow descent without stalling. It is obvious to me that, somehow, the wing-flapping of birds does, in fact, produce a very large lifting force, clearly large enough to exceed the total weight of the bird.

But how does it do that? What is there in the flapping motion that can produce a large net lift? Obviously, the downward wing motion must, somehow, produce much more lift than the opposite force that results from the wing moving upward. Actually, when I listen to the wing-flapping of a large bird, the sound produced by the down motion is louder than the sound caused by the up motion. Also, the size of the muscles which produce downward motion of the wings is much larger than the muscles which produce upward motion. In many birds, the down muscles are ten times as large as the up muscles.

I considered many possible answers to the above question. Possibly the area of the wing as it moves downward is larger than the area when it goes upward, due to a difference in wing shape. Possibly the wing is concave when going down, but convex when moving up. I considered many such possible solutions, but my engineering analysis of each proved that none of these concepts could explain how the wings could produce a large lift as a result of wing-flapping.

Finally, I discovered the answer! In fact a bird showed me the answer. As I was walking along the gravel road on my home property, a bird fell out of a tree right in front of me. It was dead. But it was still warm and very flexible. I picked it up and examined its wings. There I discovered what I believe is one of the most clever and remarkable designs in all of the animal kingdom. And, apparently, few other engineers or scientists have discovered what this bird taught me. I will tell you more about this bird in Chapter 24. The feathers of the wing are designed in such a way that when the air pressure is upward under the wing, the feathers lay flat against each other and a continuous and relatively air-tight flat surface is presented to the pressure, so that a large upward lift can result from a downward motion of the wing. Then, when the wing is moved upward, each of the feathers rotates, like a Venetian blind, which produces a large gap between adjacent feathers, through which air can move without much resistance. As the wing is moved

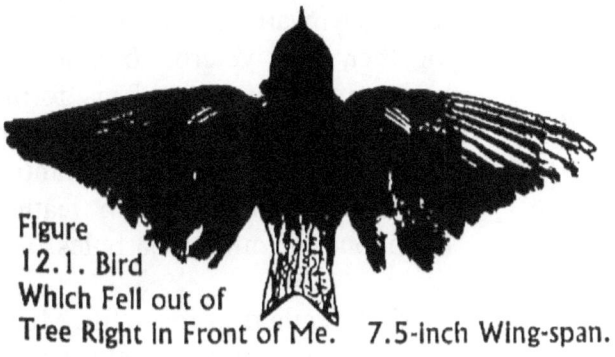

Figure 12.1. Bird Which Fell out of Tree Right In Front of Me. 7.5-inch Wing-span.

125

upward, air flows freely through these gaps and very little downward force is generated. I could blow on the upper side of the wing of this bird and the feathers would all open up the gaps. Blowing on the lower side closed the gaps and I could feel the greater force caused by the blowing. The wings acted as one-way valves. I placed the bird on my computer scanner. Figure 12.1 shows the resulting picture.

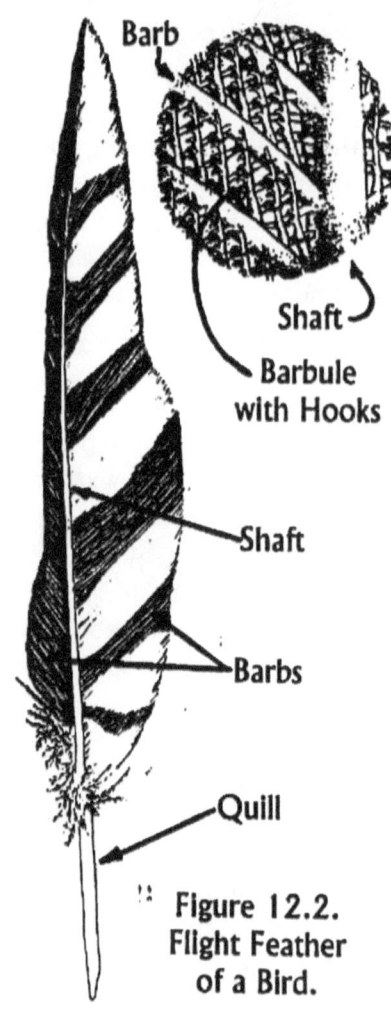

Barb

Shaft

Barbule with Hooks

Shaft

Barbs

Quill

Figure 12.2. Flight Feather of a Bird.

To explain in greater detail how the wing of a bird can function as a one-way valve, and produce net lift, it is necessary to describe the details of the design of feathers. No animals other than birds have feathers. And the key feature of the flight feathers, the secret of their success, is the fact that the shaft is substantially off center, as shown in Figure 12.2. These feathers are not biaxially symmetrical. Each feather has a shaft, which is an extension of the quill, and out from the shaft on each side are several hundred barbs. The barbs are longer on one side than on the other, thus producing the off-center shaft. Each barb, in turn, is like a miniature feather. It has, on each side, about 600 barbules, which have hooks on them. The hooks on one side face upward, while the hooks on the other side face downward. These hooks link together, like Velcro, to hold the barbs firmly in place. This forms a strong flat vane, which can hold its shape even against substantial aerodynamic forces. Each feather has about one million barbules.

Now, let's see how these off-center-shaft feathers produce lift. Much of the lift is produced by the secondary feathers which are between the bird's body and the outboard primary feathers, as shown in Figure 12.4. Figure 12.3(a) shows the feathers as they would be oriented during a downward motion of a wing. Note that they are closed, and the shaft of one feather supports the barbs of the feather next to it. The feathers are designed to be just the right distance apart so that about one quarter of the length of the barbs extends beyond the shaft of the next feather. The vane holds air pressure and a strong lift force is generated.

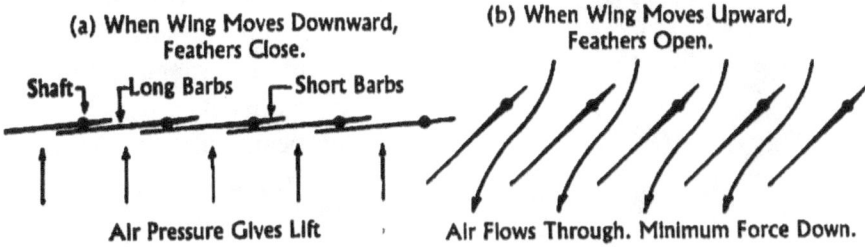

(a) When Wing Moves Downward, Feathers Close.   **(b) When Wing Moves Upward, Feathers Open.**

Shaft   Long Barbs   Short Barbs

Air Pressure Gives Lift   Air Flows Through. Minimum Force Down.

Figure 12.3 How Flapping Wing Gets Lift

Figure 12.3(b) shows the orientation of the feathers during an up stroke of the wing. The slight pressure above the wing causes each feather to rotate about its shaft. This produces large gaps between the feathers, through which the air can pass, so that little downward force is developed as the wings move upward in preparation for the next lift-producing down stroke. Now you know the secret of how birds fly. The flapping of wings, with properly placed off-center-shaft feathers does indeed produce large lift.

Figure 12.4 Widely Spaced Primary Feathers of Eagles.

Many birds are designed to use off-center-shaft feathers for another purpose, forward propulsion. Most birds have ten primary feathers, located at the outside extremity of the wings. In some birds these feathers are quite far apart, as shown in Figure 12.4. Each of these feathers has an off-center shaft, and as a result, it will be caused to rotate about its shaft when air pressure against it is developed as the wing beats up and down. As shown in Figure 12.5(a), when the wing moves downward, an upward air pressure is created below the feathers. Since each feather has an off-center shaft, the feathers will rotate as shown, the air flow will be deflected backward, and this will produce a forward thrust on these primary feathers. Figure 12.5(b) shows how a forward force will be generated, similarly, by an upward-moving wing. The feathers will rotate in the opposite direction and downward-flowing air will produce a forward force on these primary feathers. Thus, forward propulsion is produced during both up strokes and down strokes of the wing. Some birds have feathers which, in addition to having off-center shafts, also have a curvature to the shaft itself, which makes the outer portion of the feathers further off-center with respect to the center-line of the base of the shaft. This curvature provides for a larger torque to twist the feathers when pressurized by air. The wings also have other smaller muscles which can be used to control the angle-of-attack of the wings to the air, and which, with the bird's tail, enable the bird to control its motions through the air.

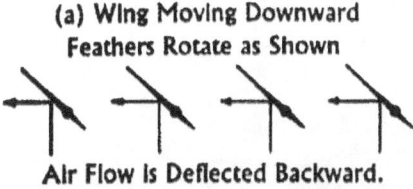
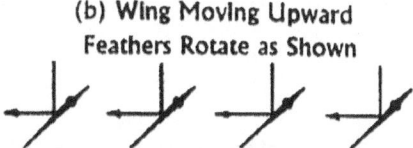

(a) Wing Moving Downward
Feathers Rotate as Shown

Air Flow Is Deflected Backward.
Feathers are Given Forward Force.

(b) Wing Moving Upward
Feathers Rotate as Shown

Air Flow Is Deflected Backward.
Feathers are Given Forward Force.

Figure 12.5 Forward Propulsion Is Provided by Widely-Spaced Feathers.

Another extremely clever mechanism that has been employed in the design of birds is the complex of muscles, tendons and bones which moves the wings up and down. This is shown in mechanical schematic form in Figure 12.6(a) and in true form in Figure 12.6(b). To accommodate the huge breast muscle which powers the down stroke of the wing, the bird has a large keelbone attached to its breastbone. To this are attached both the down-stroke muscles and the up-stroke muscles. However, the down-stroke muscle on each side is attached directly, by a tendon, to the main bone of the wing, the humerus bone. But the tendon of the up-stroke muscle goes over a pulley and then downward before it is attached to the humerus bone. It is just like a rope-and-pulley mechanism used in lifting machinery. Figures 12.6(a) and 12.6(b) show how these parts operate.

In order for birds to fly they not only must have uniquely designed wings and wing-powering mechanisms, but they must be light in weight. Their design is replete with many unique features, all of which serve the purpose of producing a flying machine that is light in weight. The bones are hollow and thin-walled, but they are reinforced with internal braces that prevent mechanical failure by elastic instability. They digest their food rapidly and excrete it quickly, to lighten their digestive systems. Their lungs are supplemented with air sacs to provide the high-capacity respiratory system needed for flight. Their hearts have very high power-to-weight ratios, due to high operating speeds. Any machine that operates at high speed can produce more power for a given amount of weight than a slower-speed machine. Many high-speed four-cycle Honda motors have taken advantage of this engineering principle. The heart of a black-capped chickadee beats at 500 times per minute when asleep, and 1000 beats per minute while in flight. Birds do not carry their young in their bodies like land mammals do, nor do they carry large numbers of eggs in their bodies as do fishes; birds generate one egg at a time and lay it as soon as it is produced.

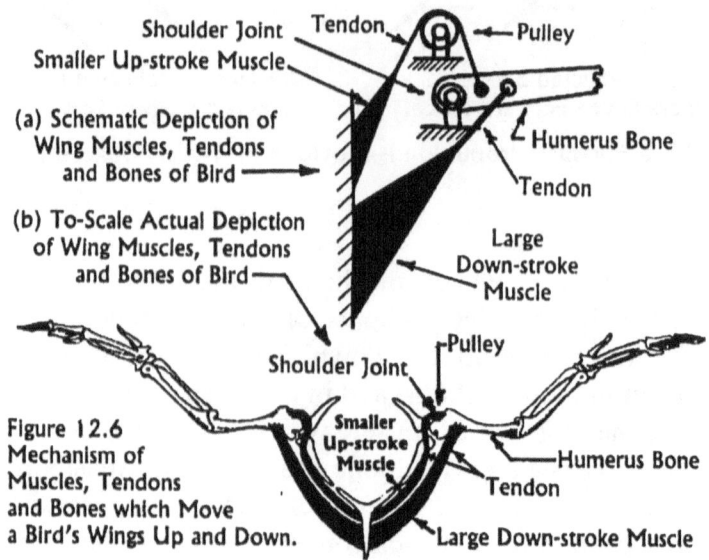

Figure 12.6
Mechanism of
Muscles, Tendons
and Bones which Move
a Bird's Wings Up and Down.

Birds also have a unique system for perching on branches which utilizes the weight of the bird's body to pull on certain tendons which, in turn, force the bird's feet to clasp the branch tightly without the need for any muscles to pull on these tendons. A bird can go to sleep and this clever clasping mechanism keeps it firmly attached to its perch. It uses the force of gravity rather than the force of muscles.

I could go on for many more pages describing the remarkable design features of birds, but I will resist that temptation so we can cover other subjects. But every athlete who has ever succeeded in lifting himself a few feet above the earth by pole vaulting, and all of the men who have tried to fly by attaching plywood sheets to their arms, will testify that any animal-powered machine that can fly in the air is, indeed, a remarkable creature.

How Bees and Other Insects Fly

Next, let's consider how insects, including bees, fly. Most insects can fly. Most insects have two pairs of wings, but, for the purpose of clarity and simplicity, we will analyze only one pair of wings. The wing design, the wing motions, and the aerodynamics of insects' wings are entirely different from those of birds. Also, the machinery which powers the wings of insects is completely different from the machinery of birds. Insects' wings do not contain any feathers. Insects' wings are unbroken flat surfaces, and their methods for developing lift and propulsion are totally different from the methods used by birds.

Let's first study the machinery which powers the wings, the muscles and the mechanisms which cause the wings to move back and forth. Figure 12.7 shows a three-dimensional view of a segment of the thorax of an insect to which one set of wings is attached. The side plates and base plate of the thorax provide a U-shaped container which houses the muscles. The wings are attached to the upper edges of this container by flexible tissues which act as hinges between the wings and the U-shaped thorax below. These hinges are labeled, P, in the Figure. Slightly inward from this line of attachment, the wings are attached by another flexible hinge, Q, to the upper plate of the thorax. It should be evident that if the top plate of the thorax were to move up and down, this would cause the wings to rotate up and down, being pivoted at the hinges, P.

These motions and muscle actions might be described more clearly with the aid of the schematic diagrams of Figures 12.8 and 12.9. Insects have a pair of muscles to move the wings upward, and an entirely different pair of muscles to move the wings downward. Figure 12.8 shows the actions of the two vertical muscles which move the wings upward. Figure 12.8(a) shows the vertical muscles relaxed. The top plate is at its uppermost position, and the wings are in their down position. Then, the brain of the insect tells the vertical muscles to contract and pull the top plate downward. The wings then rotate about the pivot points, P, and, since their inner ends have been pulled down, the wings move up to their uppermost position, as shown in Figure 12.8(b).

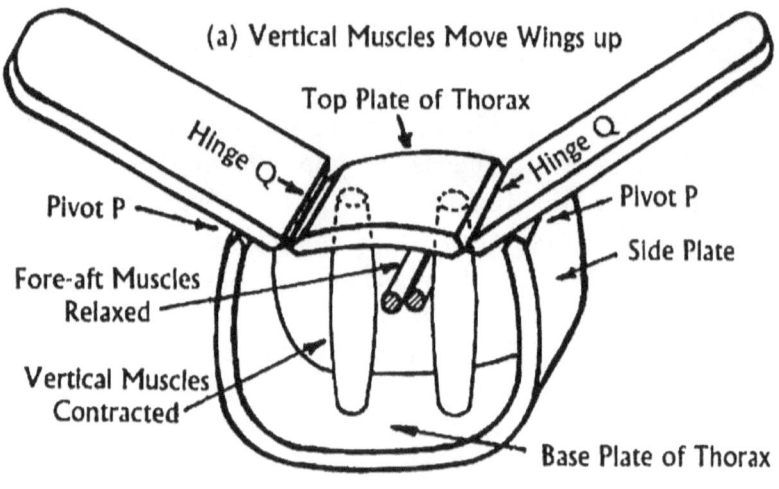

(a) Vertical Muscles Move Wings up

Top Plate of Thorax

Hinge Q

Hinge Q

Pivot P

Pivot P

Side Plate

Fore-aft Muscles Relaxed

Vertical Muscles Contracted

Base Plate of Thorax

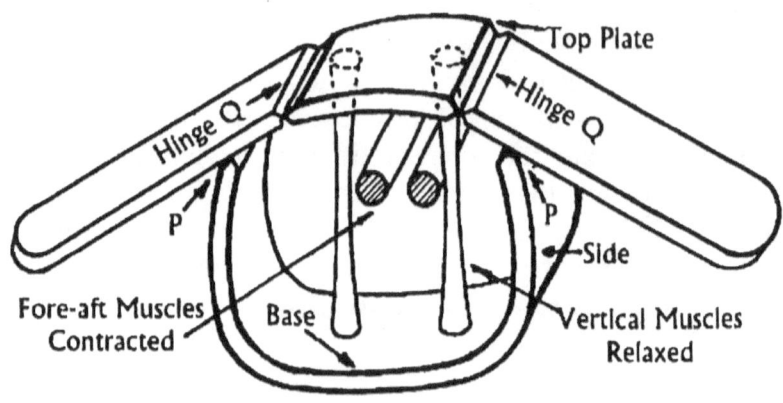

(b) For-aft Muscles Move Wings Down

Top Plate

Hinge Q

Hinge Q

P

P

Side

Fore-aft Muscles Contracted

Base

Vertical Muscles Relaxed

Figure 12.7 Three-dimensional View of Insect's Thorax Showing Vertical Muscles and Fore-aft Muscles which Move Wings Up and Down.

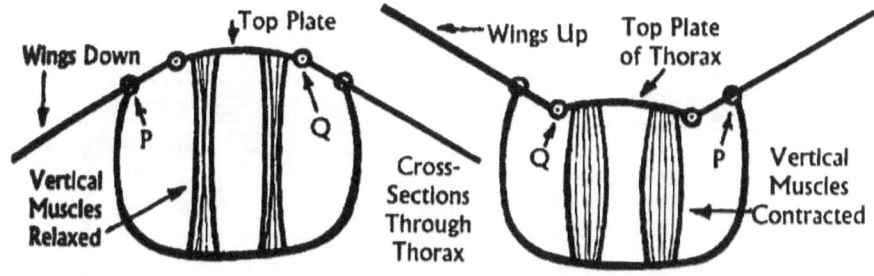

(a) Muscles Relaxed. Wings Down.   (b) Muscles Contracted. Wings Move Up.

**Figure 12.8 Schematic Diagrams of Muscles
Which Move Insect's Wings Upward.**

The fore-aft muscles are responsible for powering the wings downward. Figure 12.9(a) shows a side view of the fore-aft muscles and the top of the thorax to which they are attached. Figure 12.9 shows a longer fore-aft section of the thorax than is shown in Figure 12.7. Note where the wings are located in Figure 12.9. These fore-aft muscles basically bend the top plate of the thorax, and cause its central portion to move upward. Figure 12.9(a) shows the top of the thorax when it is bent only slightly concave downward. The fore-aft muscles are attached near the ends of this segment of the top plate. Then, the insect's brain tells the fore-aft muscles to contract, and the result is shown in Figure 12.9(b). The top plate buckles upward, and it carries with it the segment to which is attached the inner portions of the wings. This upward movement of the top plate causes the wings to move downward as they pivot about point, P, as shown in Figure 12.9(b), and also in Figure 12.7(b).

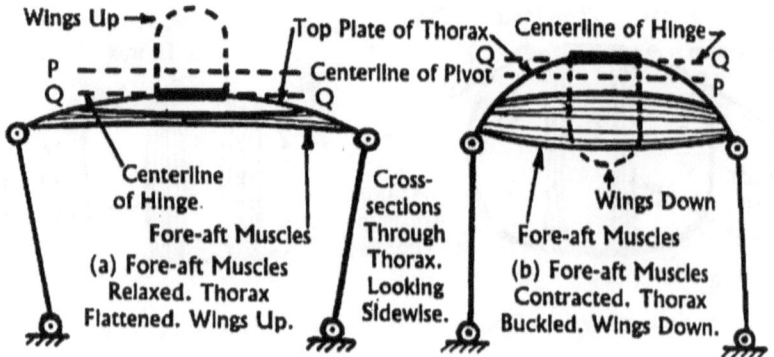

**Figure 12.9 Schematic Diagrams of Mechanisms
Which Move Insect's Wings Downward.**

Thus, by alternately stimulating the vertical muscles, and then the fore-aft muscles, the wings move up and down, or backward and forward, depending on the orientation of the body of the insect. You should recall that muscles can only generate force by contraction. They cannot push on their tendons; they can only pull, by contraction. When the vertical muscles contract, the wings move upward. When the fore-aft muscles contract, the wings move downward. If you have found it difficult to follow this description of the flight machinery of an insect, then read it again, or read it several times. Then you should marvel at the ingenuity of whoever was responsible for designing this very complicated system of muscles, tendons, hinges and the plates to which they are attached.

You should now recall that I told you that insects' wings have no feathers, and that they are continuous, unbroken, flat sheets. You should also remember that all the men who tried to fly by flapping plywood sheets fell on their noses. How, then, do insects fly, using these flat sheets for wings? They do it as follows. First, the wings of insects do not move squarely up and down. They move at an angle with respect to their bodies which is somewhat different from 90 degrees. The wings move upward-and-backward, and then downward-and-forward. Also, when an insect flies, its body is not horizontal, it is tilted upward at front, often at 45 degrees to the horizontal. As a result of these two factors, the wings of an insect beat largely back and forth, rather than up and down. Next, we should understand that the wings of an insect are quite stiff and rigid in the forepart, which is the part that, primarily, is powered, and, behind the forepart, the wings are somewhat flexible. They can and do bend. Thus, as the forepart of a wing is powered forward, the aft part bends backward and this transforms the wing into a propeller-blade exactly like that of a helicopter, and these motions give lift to the insect. Then, when the stiff forepart of the wing is moved backward, the after-part of the wing bends in the opposite direction, and it acts like a helicopter blade again, with more lift generated. Thus, lift is produced no matter which way the wings are moved.

To generate forward propulsion, all the insect has to do is tilt the wings to a slightly up-down orientation, rather than a strictly back-and-forth motion, and, just like a helicopter, forward motion will be achieved. There are, of course, other smaller muscles in the wing systems of insects, as there are in birds, to enable them to engage in the phenomenal acrobatics that we see them perform, but we will be content to emphasize primarily the basic machinery which enables them to generate lift and propulsion, and fly through the air.

Another remarkable design characteristic of the flight machinery of insects is the fact that they employ the principles of vibration dynamics to enhance the efficiency of their flight systems. Please refer either to Figure 12.7 or Figure 12.8 and note the U-shaped segment of the insect's thorax. This segment is reinforced inside the thorax and it acts like a U-shaped elastic spring. As the wings move up and down or back and forth, the upper ends of this U-shaped spring move closer together and then further apart. They would be the furthest apart when the wings are at mid-position, horizontal as shown in these Figures. They would be closest together when the wings are either at their uppermost position, or at the lowest position. The wings, and their associated parts, of course, have mass. So we have here what is known as a spring-mass system, which can vibrate; and all such systems have a natural frequency. The U-shaped thorax acts exactly like a tuning-fork, with a fixed built-in pitch, or natural frequency. Now, all the insect has to do to take advantage of this natural tuning-fork is to have its brain time the impulses which tell the wings to go up and down, or back and forth, such that the wings vibrate at the same frequency as the natural frequency of their spring-mass system. And this is exactly what they do. This minimizes the energy required for flight. What a remarkable design! If you hear the whine of a mosquito nearby, its pitch is constant, and its wings are vibrating at the natural frequency of its built-in spring-mass system.

Bats.

Figure 12.10 Bat

The third group of creatures that can fly, is bats. A bat is shown in Figure 12.10. Bats are not birds or insects; they are mammals. Their wing-flapping flight machinery is more like that of birds than of insects, but they do not have feathers and their wing-aerodynamics is more like that of insects than that of birds. But the most interesting characteristic of bats is not how they fly, but their remarkable ability to navigate in the air with the aid of a very sophisticated sonar system. So, rather than explain in detail how they fly, we will devote most of our coverage of bats to their echolocation capabilities.

All bats are nocturnal, and they like to live in caves where there is total darkness. They sleep there during the daytime and come out and catch flying insects at night. They can, in effect "see" objects in front of them by using their unique system of sound-based sonar. To achieve echolocation the bat emits sound from its mouth and listens to the reflections of the sound with its ears. The sound consists of a series of short pulses of 5 to 10 milliseconds in duration. A typical pulse may vary in frequency from about 100,000 cycles per second at the beginning of the pulse to about 30,000 cycles per second at the end. Humans cannot hear sound at such high frequencies. The pulses are then repeated at a rate of from 30 per second to 40 per second. However, as the bat gets near an object the rate may increase to 50 per second.

It is believed that a bat's brain can process the reflected sounds as the bat receives them, even when flying at high speed, so that a detailed mental image of the objects in front of the bat can be formed. Humans process light received by our eyes and formulate mental images of what we see. Bats seem to be able to do the same thing using sound waves. With this capability, bats can fly at high speed at night, or in totally dark caves, and they don't collide with objects in front of them. Using sound, they can "see" these objects and avoid them.

Who Designed the Birds, Bees and Bats?

Let's not forget that the purpose of this book is to determine who caused the Cambrian Explosion, who designed and constructed the animals of this earth. As an expert witness investigating this case, or any law case, I must first determine and examine the *facts* involved in the matter. The reason for studying birds, bees and bats is because of the *fact* that they exist on this earth. Somebody designed them and constructed them.

Evolutionists believe that birds evolved from reptiles, or from a common ancestor with reptiles. Evolutionists believe that insects suddenly appeared about 300,000,000 years ago, but they don't say very much about the details of this appearance. Bats, being mammals, are believed by evolutionists to have evolved, with other mammals, about 70,000,000 years ago. But the person who is believed to have designed and constructed these animals is nobody, just the natural forces of the earth.

On the contrary, as an expert in the engineering field of machine design, it is my opinion that these remarkable, well-engineered creatures had to have a designer and a builder. They simply could not have come into existence by chance from the elements of the earth, or by modification from some other animal which came by chance from the earth without the conscious efforts of some real designer.

In Chapter 5 we identified several of the fingerprints of a designer that I use to help me formulate an opinion on the thought processes that must have gone through the mind of whoever designed a machine or an animal. You might review them at this time, and you might apply them yourself as you think about animals. One of them asserts that if I observe a particularly clever and novel design, that would clearly be patentable by the US Patent Office, I tend to become convinced that this device had to have an inventor. Do you think the use of feathers with off-center shafts to enable birds to generate lift is a clever design? I do. It is clearly patentable! Do you think that the machinery which enables insects to flap their wings is a clever design? I do. And I also marvel at the echolocation system that bats use to enable them to navigate in total darkness. I would be willing to testify in court that these clever devices could only come from an intelligent, well-educated, and very clever inventor.

Also, the birds, bees and bats present numerous examples of other fingerprints, including: multipart systems, complex mechanical, chemical and electrical systems, ordered arrays, well-engineered shapes of parts, well-selected materials, and, especially in the case of birds and insects, they demonstrate artistic patterns and colors, all of which clearly required the services of a very well qualified designer. Do you know of any natural force of the earth, physical or biological, that could by itself invent an off-center-shaft one-way-valve for a bird's wing, a quad-muscle wing-motor system for a bee, or an echolocation system for a bat? Do you know of any natural force of the earth that understands advanced aerodynamics, mechanical vibration theory, or engineering acoustics? Do you know of any natural force of the earth that can invent clever mechanisms? If so, please tell me about it and explain to me in engineering terms how it works.

And so in conclusion, my expert-witness testimony to you, the Judge and Jury, is that the birds, insects and bats are obviously so ingeniously conceived and well engineered that they had to have come into existence through the efforts of a real being who was an extremely capable design engineer. Possibly, in a later Chapter, we will be able to identify this remarkable being.

CHAPTER 13. NEW-FEATURE-PRODUCING AGENTS

In this Chapter, also, there might be a little fiction mixed in with many facts. But you will be able to tell which is which.

South Pass on the Oregon Trail

Today, just south of Lander, Wyoming, Highway 28, heading westward, climbs over the Continental Divide in the Rocky Mountains. The summit is called South Pass. In the 1830's, the Oregon Trail was blazed, from Missouri to Oregon, a distance of more than two-thousand miles. At its halfway point it too climbed over South Pass, at an elevation of about 7500 feet. In the 1840's and 1850's thousands of pioneer Americans trudged westward over South Pass, with ox-drawn wagons, to seek their fortunes in the Oregon Territory. For thousands of years prior to the 1830's Shoshone and Cheyenne Indians, and their predecessors, traveled over South Pass, on the Oregon Trail, on foot, and on horseback.

Anthill Overlooks South Pass

On a small hill overlooking South Pass there is an anthill, which has been there for thousands of years. As you know, ants are remarkable creatures, and their colonies and their civilizations are very complex, very advanced, and worthy of study by human beings. Ants live in colonies, in which hundreds, or even thousands of ants may reside. There are three classes of ants in every colony: queens which are the females, males, and workers. The worker ants specialize, each doing the work of its specialty. Some feed and care for the young. Some keep the nest clean. Others gather and store food. Some are soldiers which guard the entrances to the nest. Some ants raise other insects, as humans raise cows. They get sugars and other liquids from "ant-cows" such as aphids.

Ants Teach Evolution in their Schools

What is lesser known by humans is that ants have schools, where there is a teacher and many students. I will reveal to you what I learned was taught by the teacher-ant at South Pass on a particular day last year. The subject was evolution, particularly the evolution of the conveyances used by the humans who climbed over South Pass during the past many thousands of years. Observations had been made by the ants, and these were recorded in their library. The teacher-ant first explained that

No Saddle

Horse With Saddle

humans traveled over the pass, riding on the backs of horses, many years ago. Then one day a horse, adapting to its environment, was struck by lightning, and a growth suddenly appeared on its back. This, the humans called a saddle. Other horses soon evolved with similar saddles. Since the horses with saddles could more efficiently carry their riders, by natural-selection, the horses without saddles soon became extinct. Then, on another day, again adapting to its environment, one horse, due to a mutation or two, developed two long poles, one on each side, which were attached to the saddle, but the aft ends of the poles dragged on the ground. The humans called this a

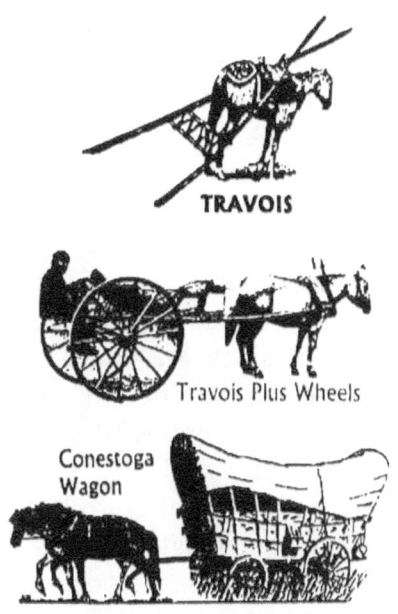

TRAVOIS

Travois Plus Wheels

Conestoga Wagon

travois, and they rode, and carried freight, on the poles. The horses without travois, being less efficient, through the action of the principle of the survival of the fittest, soon became extinct. Then, one day, by many mutations, and with the help of Mother Nature, one horse, adapting to its environment, developed wheels on the ends of the poles. Other horses followed with wheels, and this caused the travois-only horses to become extinct. Further adaptations and mutations brought forth two-wheeled horse-and-buggy conveyances, and then, in the 1830's, four-wheeled wagons evolved.

Finally, by further adaptation to the environment, and after many more mutations, and with help from teleology and cosmic rays, a horse evolved into an engine, and the horseless carriage emerged. This conveyance was so efficient that, by natural-selection, horses disappeared altogether, and the horseless carriages, over time, evolved further to become lower, wider, longer, more streamlined, and they moved at much higher speeds. Humans called these conveyances automobiles. Today automobiles by the thousands climb over South Pass at blinding speeds, just beneath our ant-hill. And, said the teacher-ant, those are the facts concerning how automobiles evolved from horses, through the agencies of natural forces, by evolution.

Oldsmobile

Parry

Chevrolet

Ford

But, said one student-ant, isn't it possible that all of these new conveyances were designed and constructed by some super-ant, or some super-being? No! That is not possible, said the teacher-ant, because no ant has ever seen such a super-being; and therefore we know that no such being exists. And, since we know that no super-being exists, we must conclude that the automobiles came into existence by evolution, by the actions of new-feature-producing agents, such as mutations, and by natural-selection. Thank you, said the student-ant. As always, we students believe what our teacher tells us.

And now, may I ask you, dear Reader, do you think it is preposterous for the ants to believe that a horse could evolve into an engine? Evolutionists believe that a lizard evolved into a bird. I think that both of these propositions are about equally preposterous. The ants have never seen the designer of an automobile, and evolutionists have never seen the designer of animals, hence the ants and the evolutionists do not believe that these designers exist. But we, as humans, know that the designers of automobiles were, in fact, living, intelligent, creative human beings. The purpose of this book is to identify who was the living, intelligent, creative designer of the animals.

Can One Animal Evolve into Another?

Since evolutionists do not believe in the existence of a supernatural designer-craftsman, and since they now know that only a few billion years ago the earth consisted only of rock, air, and water, and there were no animals, they are forced to conclude that, somehow, by chance, sea-water turned into a protozoan, and then, the protozoan evolved into all of the animals we see today. The theory of evolution must consist of two entirely separate propositions: (1) it must explain how sea-water turned into a protozoan, complete with DNA, and (2) it must explain how the protozoa evolved into other animals. The theory of mutations, followed by natural-selection, cannot apply to sea-water evolving into a protozoan because the sea-water has nothing with which to mutate. It has no cells and no DNA molecules, or anything remotely similar to DNA. We have proven, in Chapter 10, that the first step, the creation of protozoa by chance, is impossible. In this Chapter we will address the likelihood that the second step is feasible. Can one animal evolve into another?

New-Feature-Producing Agents

Evolutionists believe that protozoa evolved into the other animals by a slow process of many naturally-caused changes, followed by natural-selection. I will call the agents responsible for these naturally-caused changes, "New-Feature-Producing Agents", or NFPA's. According to evolutionists, NFPA's must be naturally-occurring agents. In previous Chapters we have identified several such agents. These included the following: wind, rain, hail, snow, flooding, freezing, thawing, earthquake, lightning, ice-movement, cosmic rays, radiation from the sun, etc. These are natural forces which some evolutionists apparently believe can create new features. This list of NFPA's was presented in connection with our study of the fire-ring of rocks on a mountain-top, and the presumed creation of protozoa from sea-water. To be meticulously scholarly on this subject, we should question whether or not this list of NFPA's is complete. Are there other NFPA's that should be added to our list? The answer to this question is, yes. Others have been suggested, and we should give consideration to all other NFPA's that have been proposed to exist. Our expanded list should probably include the following: Mother Nature, Teleology, Adaptations, Mutations, Mendelian genetics, Punctuated equilibrium, and Natural-selection, itself. Let's consider these one at a time. Actually, due to their importance, separate Chapters will be devoted to several of the above.

Mother Nature

We often hear the general public speak of "Mother Nature". This term is often applied to identify whatever force is responsible for designing the marvelous and clever features possessed by animals. "Isn't Mother Nature wonderful?" is often heard, as applied to the animals of the earth. But no detailed identification of Mother Nature is ever offered. Even many people who use the term do not really believe that any person called Mother Nature really exists. When used to explain evolution, it is obviously a convenient, purposely-imprecise term used by those who have some weakly-held belief in the existence of a supernatural designer-craftsman, but who do not wish to be categorized as among those who do not believe in the theory of evolution. They are fence-straddlers. To me, the concept of Mother Nature is vague and meaningless. It is in the same category as the Tooth-Fairy, Santa Claus, and Jack Frost. It is a term used by fuzzy-thinkers and outright deceivers. It is not to be taken seriously. Mother Nature is not a person who could design machinery, and Mother Nature is not a person who could design animals. So let's conclude that Mother Nature is not a New-Feature-Producing Agent (NFPA) that has any credibility.

Teleology

Teleology is a philosophical doctrine that asserts that purpose and design are a part of nature. This doctrine asserts that phenomena are guided not only by mechanical forces but that they also move toward certain goals of self-realization, and that ultimate causes in nature exist. In non-philosophical terms this means that nature, itself, has some ability to design and construct. To my way of thinking, teleology is little more than a belief in Mother Nature at a higher intellectual level held by philosophers. Vague and mysterious forces are not capable of designing machinery, and neither are they capable of designing animals. Every machine must be designed by an intelligent, existing being, and each animal must also be designed by an intelligent being. Teleology is not an NFPA that deserves to be taken seriously.

As stated above, in subsequent Chapters we will consider other proposed New-Feature-Producing Agents, including: Adaptations, Mutations, Mendelian genetics, Punctuated equilibrium, and, Natural-selection, alone.

CHAPTER 14. ADAPTATIONS. AN NFPA?

Can Adapting to an Environment Produce New Features?

The term, "adaptation" is widely used by evolutionists, and, for that reason it deserves to be treated as a separate Chapter. A typical textbook on the biology of animals might use the term hundreds of times. To an evolutionist the term means that an animal evolved, or changed, in such a way as to adapt to a particular environment, and the environment, itself, is believed by some to have caused such changes. The changes, or adaptations, are thought to make the animal better able to survive in a particular environment. Evolutionists look around the world and they see a wide variety of environments, such as the oceans, the air, the deserts, rain forests, the tropics, the poles, midwestern United States, etc.; and in each environment they see animals that are surviving there. From this, some evolutionists conclude that the environment itself has been a New-Feature-Producing Agent that has produced features which have enabled the animals to survive in a particular environment.

For example, let's assume that cats did not always have claws which would enable them to climb trees. Then some cats found themselves in an environment where there were trees and dogs, and the dogs preyed upon the cats. So, the cats adapted to their environment and evolved claws so they could climb trees and get away from the dogs. Other examples of the concept of adaptation might assert that cows found themselves in grasslands, so they evolved their extremely complex four-chamber stomach and cud-chewing digestive system so they could make use of the grass. The giraffe found that the leaves high up on trees were not eaten by others, so he adapted and evolved a long neck. Reptiles saw that there were insects up in the air not eaten by others, and the reptiles noted also that predators were constantly chasing them, so they adapted and evolved wings and became birds.

Can the Environment Design and Construct New Features?

Some evolutionists seem actually to believe that an environment can stimulate an animal in some mysterious way to cause it to produce new features. They believe that an environment can be a real New-Feature-Producing Agent, an NFPA. Other evolutionists do not believe this, but rather they believe that some other agents produce the new features, and the environment serves only to act as a natural-selection agency. They, therefore, do not consider the process of adaptation to be an NFPA, but it is merely another name for natural-selection.

How credible is it to believe that an environment can actually produce new features in an animal? Can the environment produce eyes, ears, claws, fingers, wings, fins or tails? Can the environment design? Can the environment construct? Can the environment design and construct machinery? Can it design an animal? Can you identify precisely what is the capability that the environment could possibly posses which would enable it to produce new features in an animal? If you can't identify this capability, isn't it reasonable to conclude that animals and machines can only be designed and constructed by real, existing, knowledgeable and creative beings, and certainly not by the environment.

The Environment has no Creative Power

If I were to throw a fish, a squirrel, and a raccoon into the middle of the ocean, the fish would survive, but the raccoon and the squirrel would die. Does that mean that the fish adapted and the others didn't? No! It merely means that the designer of the fish designed it in such a way that it could survive in the ocean. If I were to throw a bird, a rabbit and a fish into the air off of a high, steep cliff, only the bird would survive. It was designed to survive in the air.

Many evolutionists would agree that drastic changes in animals probably would not occur suddenly. But they may believe that the environment could bring about drastic changes a little bit at a time over a long period of many years. However, if the environment really has no creative power whatever, it could act forever and it would produce nothing new. I could practice jumping over my house forever, but I could never do it. It's just something that I can't do. The environment has no ability to design or construct. It can only select from among designs created by other agents. The power of natural-selection is very real. It does select the fittest, and the unfit become extinct. But the environment has no ability to design the fittest.

Adaptation is Nothing Other than Natural Selection

Engineers design submarines for deep ocean survival. Engineers design aircraft for survival in the air, and tanks for survival in the desert. They design snowmobiles for survival on snow; and buildings to survive in any kind of weather. Similarly, various animals were designed to survive in different environments on earth. And, as environments change, or animals migrate, obviously each animal that is suited for a particular environment survives in that environment, and the others die. As an end product, each animal has adapted to its environment, but only in the sense that, by natural-selection, it has survived in a particular environment and others were not equipped to do so.

Could Adaptation Produce New Features for You?

Do you think that the environment could produce new features for you? Let's assume that you lived a few thousand years ago, and the main diet for you and your neighbors in your environment was the eating of rabbits. Now, rabbits are hard to catch. So, for you, it would be very beneficial if the powers of adaptation evolved somewhere on your body two spinning facilities that would produce heavy-duty, sticky webs, like the spinners that spiders have. You could then spin webs that would catch the rabbits, and then you would have all the rabbits, and your neighbors would starve to death by the process of natural-selection. What statistical probability do you think there would be for the powers of adaptation to design and construct on your body such web-spinning capability?

Or, possibly, by adaptation, maybe you would like to have an extra eye evolve on the back of your head. Then none of your enemies could sneak up on you from behind and injure or kill you. Your chance of survival would be enhanced. Do you really think that the environment has the power to design and construct such useful new features? I know of no evidence which would suggest that it has that power.

Why Aren't All of the Animals Alike?

If adaptation to an environment had any really substantial ability to produce new features, and change one animal into another, then, in any one specific environment, all of the animals would be the same. They would have all adapted to the same environment. I have lived for many years in central Ohio; and, as I look about me I don't observe that all the animals are the same. In fact, there are probably hundreds of thousands of different species of animals in central Ohio, and they are all amazingly different one from another. We have ants, birds, mice, raccoons, foxes and deer. They clearly are not all alike. Certainly the process of adaptation has failed in Ohio. In fact it has failed everywhere to produce any new features or species. The wide variety of animals in any selected region of the earth tells us that they are not designed by adaptation.

The Term "Adaptation" is Widely Misused

Nevertheless, the term "adaptation" is used by many biologists and evolutionists as a synonym for evolution. To them, "adaptation" and "evolution" have the same meaning. It is probably true that many of our most competent, up-to-date, and fair-minded evolutionists agree that an environment cannot really design and construct new features for animals, and that the term "adaptation" is basically just another name for the process of natural-selection. But the term "adaptation" is so widely used in connection with the concept of evolution that other evolutionists, professors, teachers, and the general public are lured into the habit of using this term to imply that the environment can actually produced new features and has perfected them into useful final products. This misuse of the term is widely practiced in our schools, in our textbooks, and in our communications media.

A Paleontologist's Conclusion

The esteemed German paleontologist, Otto Schindewolf, said, "We observe that organisms are adapted . . . in accord with their environment . . . From this situation . . . it was assumed . . . that it was environmental factors that shaped the organism . . . the driving forces of morphological transformation . . . Currently, . . . this . . . has been almost completely abandoned, for the naive assumption that evolution is affected by . . . adaptations . . . is in conflict with the results of genetics . . . the types could not have arisen by adaptation and selection." (Schindewolf 1)

Conclusion: Adaptation is Just Another Name for Natural Selection

We must conclude, then, that the process of adapting to an environment has no ability to produce new features. Adaptation is merely another name for natural-selection.

"Adaptation" is certainly not a New-Feature-Producing Agent. It is not an NFPA.

The Cambrian Explosion

CHAPTER 15. MUTATIONS. AN NFPA?

Of all the New-Feature-Producing Agents proposed by evolutionists the one that is most often mentioned is mutations. Hence we will devote a very carefully detailed Chapter to the study of mutations.

Evolution Needs a Valid NFPA

Let's review again the role of an NFPA in any theory of evolution. There must be no misunderstanding of its importance. In order for the theory of evolution to explain how all the animals of the earth could have evolved from the protozoa, it is necessary to demonstrate that there is at least one New-Feature-Producing Agent (NFPA) that is clearly functional. A valid NFPA, plus natural-selection, are the two ingredients needed to produce evolution. Natural-selection only acts to cause the extinction of animals which are not fit. So, the validity of evolution really boils down to the search for a credible NFPA.

Mutations Plus Natural Selection are Proposed

For many years early evolutionists thought that new species could be created by the process of use or disuse of parts and capabilities, followed by the inheritance of acquired characteristics. Scientists later proved that this theory was contrary to the laws of genetics. So it has been abandoned. In the recent past many high-school teachers, college professors, textbook writers, and other evolutionists have been teaching that evolution is a fact, and that its method of operation is a combination of mutations plus natural-selection. So let's now consider mutations.

DNA Atoms are the Blueprints for Animals

In order to explain what mutations are, and how they are presumed to work, we need to understand DNA. To learn about DNA we might first study its analogue in the realm of machinery. Figure 15.1 shows a drawing of the crankshaft of the engine of a modern automobile. This drawing contains about 300 bits of information, and it is for just one part of an automobile. There are well over 5,000 parts in an automobile, so you can see that hundreds of thousands of bits of information are needed in the plans for an automobile before it can be constructed. Hundreds of drawings are needed; but, once completed, all of the information required to make an automobile is contained in the drawings. The drawings are made by engineers; and then the auto-workers in a factory follow the drawings and make the car. Similarly, every animal has a set of plans, a set of drawings. The blueprints for an animal, its drawings, are its DNA.

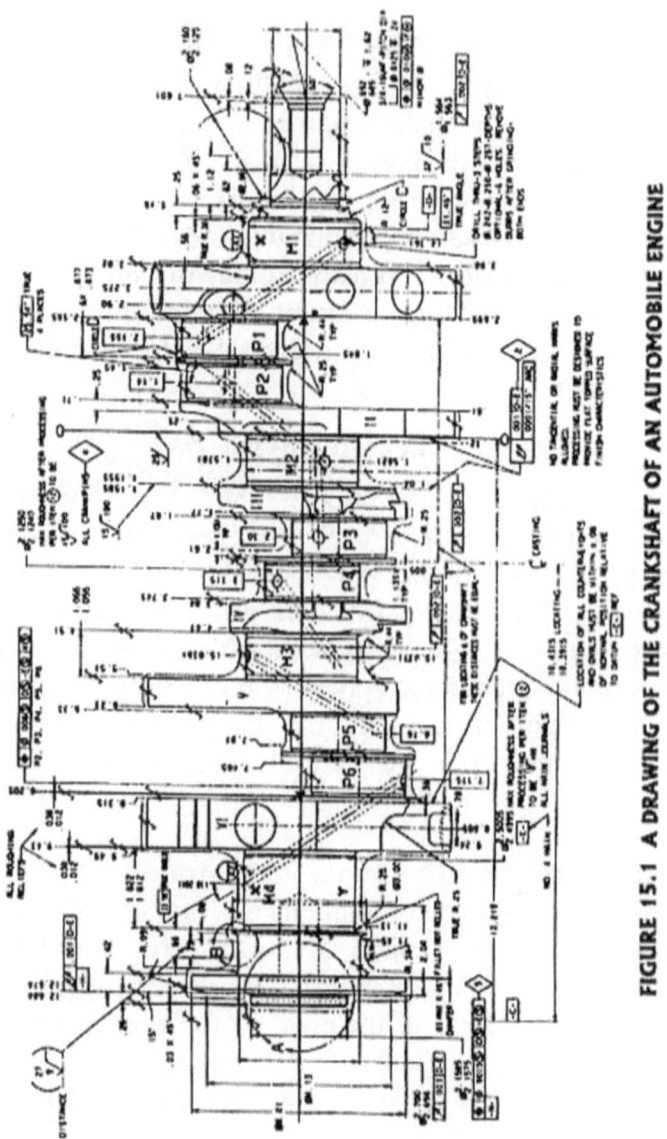

FIGURE 15.1 A DRAWING OF THE CRANKSHAFT OF AN AUTOMOBILE ENGINE

What is DNA?

The term, DNA, is an abbreviation for deoxyribonucleic acid. DNA is an extremely complicated molecule, containing millions of atoms. Genes are specific segments of the DNA molecule. Every cell in every animal contains a DNA molecule, and that molecule directs the structure and function of that cell. The DNA is like a book of instructions in every cell, and for the animal as a whole. The DNA molecule is different for every animal species, and it is different for every individual animal. It determines the specific structure for every part of the animal, and it manages the construction and function of every organ. It determines the personality and instincts of the animal, and it guides its growth and decline, from conception to death. What do these marvelous molecules look like?

The Chemistry of DNA

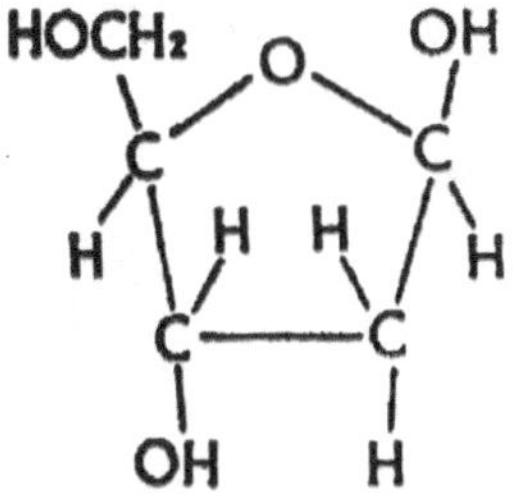

Deoxyribose Sugar (S)
Figure 15.2

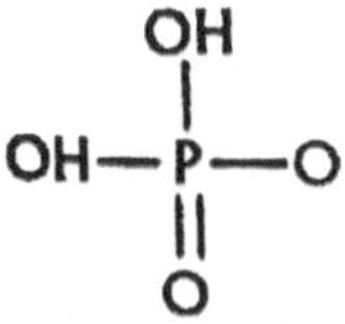

Phosphate Group (P)
Figure 15.3

Every DNA molecule consists of a specific assemblage of many nucleotides. A nucleotide is an aggregation of three constituents: (1) a 5-carbon sugar group, (2) a phosphate group, and (3) a nitrogen-containing chemical base. Each of these constituents is composed of a particular group of atoms. Figure 15.2 shows the sugar group. It is called deoxyribose. We will refer to this group of hydrogen, carbon, and oxygen atoms as sugar, S. Figure 15.3 shows the phosphate group. We will call it P. It contains hydrogen, carbon, oxygen and phosphorous. There are four bases in DNA: Adenine, Guanine, Thymine, and Cytosine, which we will call, respectively, A, G, T, and C. Figures 15.4, 15.5, 15.6, and 15.7 show these nitrogen-containing bases.

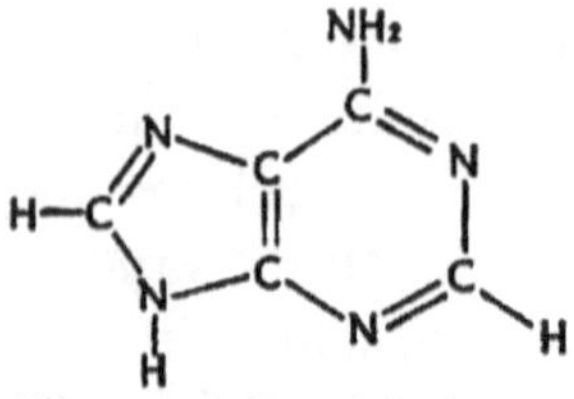

Nitrogenous Base, Adenine (A)
Figure 15.4

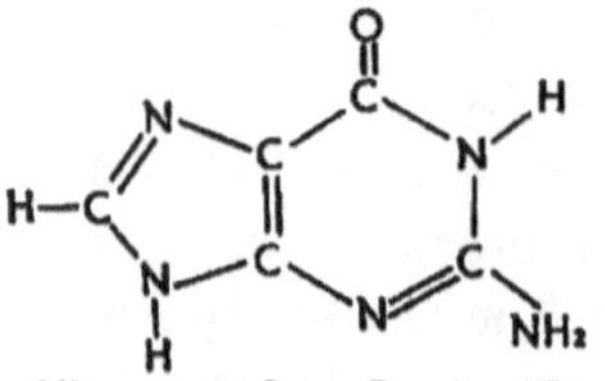

Nitrogenous Base, Guanine (G)
Figure 15.5

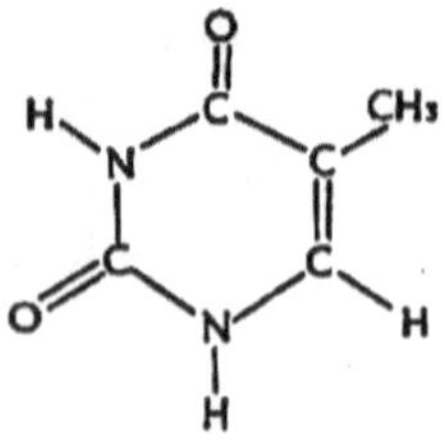

Nitrogenous Base,
Thymine (T)
Figure 15.6

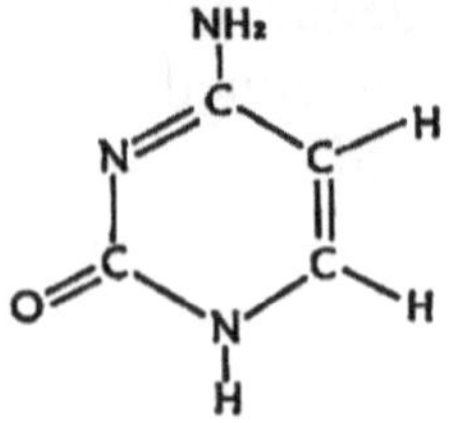

Nitrogenous Base,
Cytosine (C)
Figure 15.7

The DNA molecule consists of a ladder-shaped assemblage of these components, arranged as shown in Figure 15.8. The ladder has two upright members and many connecting rungs. Each upright is a chain of alternating sugar groups and phosphate groups. Each sugar group is attached to one of the four bases, and the bases of the two uprights are then paired and joined together to form the rungs of the ladder. Since Adenine and Guanine are larger than Thymine and Cytosine, to ensure that all of the rungs have the same width, and hence fit well together, Adenine always pairs with Thymine, and Guanine pairs with Cytosine. Finally, the ladder structure is twisted into a double helix, as shown in Figure 15.9. The whole DNA structure may be further twisted, wound, coiled and compacted to save space.

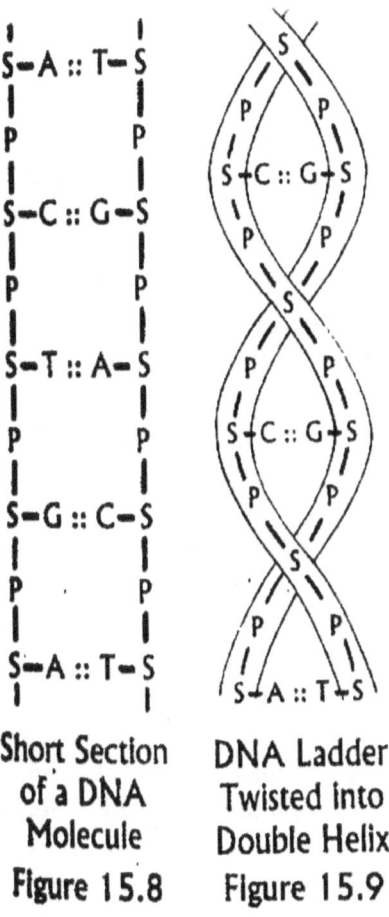

One molecule of DNA for a human being contains billions of atoms, and these atoms are arranged in a pattern that is specific to each individual. The primary distinctive feature of a particular DNA molecule is the pattern of its sequence of base-pairs in the ladder. It has been estimated that the DNA molecule in a human cell, if untwisted and unwound, would be about 13 feet in length, but when twisted, coiled and packaged in a cell it may have a length of only 0.008 inches. The most significant essence of a DNA molecule is not its chemistry, but rather the fact that it is a repository of information! Its information is contained in the pattern of its almost infinitely complex arrangement of atoms, and this information is then used to direct the development and life of the specific animal which corresponds to a particular DNA molecule.

Short Section of a DNA Molecule

Figure 15.8

DNA Ladder Twisted into Double Helix

Figure 15.9

A New Feature Could Only Result From a Change in the DNA

My purpose in explaining in some detail the general nature of DNA is to enable you to appreciate its fantastic complexity, and as a result you can better understand that these DNA molecules are, in fact, the blueprints, drawings, and plans which determine in every detail how each animal will be constructed and will function. And, if one animal is to evolve into another animal, it will have to do so by effecting changes in the first animal's DNA, possibly due to the effects of mutations. So, we must consider whether or not mutations could possibly be the New-Feature-Producing Agents which, with natural-selection, could cause one animal to evolve into another.

What are Mutations?

A mutation is a sudden change in the DNA of an animal which will persist and be inherited. Mutations can be caused by such items as the following: bombardment by radiation from the sun, X-rays, cosmic rays, viruses, chemicals, and by errors in DNA replication. To understand the basic concept of how a mutation might occur, we should visualize a DNA molecule and then imagine how this molecule could be changed. Each DNA molecule of a human being contains about 3 billion pairs of nucleotides arranged in a double-helix ladder. If you count the number of atoms in a single nucleotide, based on the chemical representations of Figure 5.5 and Figures 15.2 through 15.7, you will determine that each nucleotide contains about 34 atoms. Hence, we may conclude that a single DNA molecule in a cell of a human being would contain the following number of atoms:

$$A = (3,000,000,000)(34)(2) = 204 \text{ billion atoms.}$$

The information contained in this complex of atoms constitutes the drawings and plans which determine all of the structures and functions of the cells of the individual. Now visualize some X-rays, cosmic rays, or radiation from the sun bombarding this molecule. Such rays could displace, remove, or replace some of these atoms of the DNA molecule. If this happened, the drawings would be changed, and new features would be produced in the individual. Or, as the cells of an individual divide to produce new cells, and the DNA molecule must be copied for each new cell, there could be an error in copying, so that the DNA of a new cell would be different from that of the mother cell. This would produce new features. There is no question but that mutations, due to radiation bombardments or copying errors do occur. The wings of fruit flies have been changed by exposing them to X-rays. Human beings are now believed to be subject to more than 3500 mutational disorders, but, since we have two sets of genes, these diseases rarely come to the fore.

Mutations are not Likely to Produce Useful New Features

So, mutations do occur, but they are not likely to be the NFPA's that are needed for the theory of evolution, for the following four reasons: (1) mutations occur only very rarely, (2) almost all mutations are harmful, not beneficial, (3) to produce new features, mutations must, somehow, add new atoms to the DNA, and (4) a long series of many related beneficial mutations would have to take place to create any improved features that could be acted upon favorably by natural-selection. We will consider these four reasons, one at a time.

Mutations are Rare

Mutations occur only with great rarity. When the DNA of one cell is replicated, it is not only copied but enzymes actually perform the function of proofreading the replication. Consequently, it has been estimated that there will be only one replication error in the copying of 100,000,000 nucleotides, which would contain more than 3 billion atoms. (Starr 1) Another estimate, which includes mutations caused by radiation, replication and other sources, concludes that only one individual animal in a million will experience a mutation. (Ross 1) Mutations, indeed, are very rare occurrences.

Most Mutations are Harmful

Consider, next, the ratio of harmful to beneficial mutations. Almost all known mutations are harmful. Estimates of the ratio of harmful to beneficial mutations range from 10,000 to one, to 1,000,000 to one. (Ross 1) In spite of the fact that, for the past 50 years, evolutionists have touted beneficial mutations as the primary source of new features, and the real engine of evolution, most evolutionists could not name ten confirmed examples of beneficial mutations. Many could not name one. Ask your evolutionist friend to name a few examples of beneficial mutations. If the theory of evolution were factual, thousands of examples of beneficial mutations should have been documented during the past 100 years. I suspect your friend cannot name one. Sickle-cell anemia is often cited as an example, but really this result of mutations is beneficial only in that it reduces deaths from malaria. If the major proponents of evolution cannot cite dozens of clearly documented instances of evolution's primary alleged method of operation, then must we not conclude that evolution is clearly a theory, not a proven fact?

Do you think it would be wise for you to spend hours near an X-ray machine, or inside of a nuclear power plant? Would it be wise for you to go to Chernobyl, in Russia, where a nuclear power plant exploded? Should we actively try to destroy the ozone layer that shields us from radiation? If such radiations are likely to cause you to evolve, and develop new beneficial features, then you should seek to be bombarded as much as possible by these sources of radiation. Maybe you could get a new eye in the back of your head. In reality, if you are smart, you will avoid such radiations, because they are much more likely to damage you than to improve you.

Being bombarded by mutation-causing radiation, would be like shooting a new car with a 30-caliber rifle. Let's assume that it would be beneficial if the ballast resister in your ignition system were located inside the interior of your car, under the dashboard, rather than out near the hot engine. To move the resister, it would be necessary to produce a hole of 0.30-inch diameter in the bulkhead between the engine and the car's interior, through which wires could pass. This hole could be produced by shooting the car with a 30-caliber rifle. However, following the mutation analogy, there must be a minimum of 10,000 harmful mutations for every beneficial one. So you must shoot the car 10,000 times if you want to get one shot that would produce a hole where needed to move the resister. You must walk around the car several hundred times until you have shot it with 10,000 bullets. What would be the result? The car would have no tires, no windows, no gas tank, no radiator, no heater, no air-conditioning, no usable engine, etc. It would be totally demolished.

Similarly, it would be highly unlikely that mutations would do anything other than damage you or an animal. Mutations caused by DNA copying errors would have a similar result. The billions of atoms of the DNA molecule, the plans for an animal, already contain what is needed to produce a marvelous machine. The animal's DNA is not likely to be improved by mutations. Mutations are harmful by a ratio of at least 10,000 to one. Radiation and copying errors do not produce new features that are beneficial.

Atom-Adding Mutations

The third reason that mutations are not likely to be the New-Feature-Producing Agents that are needed by the theory of evolution is that, in order to create new features, many atoms must be added to the DNA of an animal to produce a new feature, or a new species. We will call this process of producing a mutant, the atom-adding type of mutation.

It has been estimated that there are about 3,000 times as many atoms in the human DNA as there are in the DNA of the E. Coli Bacterium. There are probably about the same number of atoms in the DNA of a single-celled protozoan as in the DNA of an E. Coli Bacterium. You may recall that the human DNA contains about 204 billion atoms. Thus, to evolve from a protozoan to a human being would require the addition of well over 200 billion atoms to the DNA molecule of the protozoan. Presumably, animals intermediate in complexity between a protozoan and a human would have DNA molecules with intermediate numbers of atoms. So, to go from a lizard to a bird might require the addition of only one or two billion atoms.

The atoms that would need to be added would include carbon, hydrogen, oxygen, nitrogen, and phosphorus. Carbon is found in the air, in the form of carbon-dioxide. Nitrogen comes from the air. Hydrogen and oxygen are the components of water. And phosphorus comes from the ground. The problem is not just a matter of finding where these atoms exist, but they must be removed from where they exist and be placed in the DNA molecules in just the right places. They must be so organized as to form the fantastically complex sugar groups, phosphate groups, and nitrogenous bases shown in Figures 15.2 through 15.7, and they must be positioned in the twisted double-helix ladder in just the right positions to operate as part of the DNA molecule. Furthermore, to have any lasting effect on offspring, these billions of atoms would have to seek out and find the particular gamete cell that is going to produce the next progeny. Could this happen by chance? Do you really believe that this transportation and positioning of atoms could occur by chance? What could cause these atoms to float through the air from the land, sea, and air and go to the DNA molecule of an animal?

In the development of automobiles, as the years have rolled by, simple cars have been improved to become more complex. In 1901, the steering wheel, instead of a tiller, was introduced. Also, water-cooling and detachable tires were first used in 1901. In 1902, Packard patented the H-slot gearshift. In 1906, Buick included a storage battery. In 1908, a four-wheel-drive car was introduced. In 1911, Marmon introduced the first rear-view mirror. In 1912, Kettering developed an electric starter. In 1918, four-wheel brakes were introduced. In 1926, Packard offered hypoid gears in the differential. In 1928, Cadillac introduced a synchromesh transmission. In 1939, Oldsmobile offered a Hydra-matic transmission. In 1953 most cars began to use a 12-volt electrical system.

These changes did not happen by chance. These cars did not mutate. These innovations did not originate due to errors in copying the drawings, or by radiation bombardments. Every one of these changes originated through the design efforts of engineers, human beings; and new drawings were made by these engineers, from which the new cars were built. In order to add new features to automobiles, new drawings had to be added, not subtracted. Copying errors or bombardments that would damage the drawings could only subtract from the existing drawings. New features could only come from the addition of new drawings designed by human beings.

The drawings for animals are the DNA molecules. These molecules are almost infinitely more complex than the drawings for automobiles. Copying errors and bombardments by radiation could only subtract from the existing DNA molecules. New features could only come from adding to the DNA, not subtracting from it. If you agree that changes in automobiles had to come about by the efforts of engineering designers, how could you believe that new features could be produced in animals by copying errors and radiation bombardments? Or, realizing that new features could only be produced by adding new atoms to the existing DNAs, how could you believe that this could come about merely by chance? For the creation of every new more-complex animal from a simpler animal, thousands of new atoms had to be added to the DNA of the simpler animal. Can you explain how these added atoms could be transported from the earth's environment to the DNA molecules of an animal? Atom-adding mutations could not possibly produce new features, or new species, by chance.

Multiple Related Mutations

The fourth reason that mutations could not possibly be the New-Feature-Producing Agent that evolution needs is because each mutation, at best, can only produce a minor change; but a new feature, a new organ, a new part, or a new species obviously requires hundreds or thousands of minor changes, before it could be treated favorably by natural-selection. Any partly-finished, non-functional new part would quickly be eliminated by natural-selection. Only completely finished, fully functional parts would be allowed to persist. To create a viable new feature it would be necessary to have many mutations, and they must all be related to the same objective. We will call this the concept of multiple related mutations. Let's consider an example. Evolutionists believe that all animals evolved from the simplest animal, the protozoan. Protozoa do not have eyes, as do the vertebrate animals, such as mammals, birds, reptiles, and fishes, including humans. Let's explore the statistical probability that the vertebrate eye shown in Figure 15.10 could have been produced by mutations and natural-selection. (Hickman 3) The first event that would have to take place in order to have eyes evolve in an animal's head would be to experience errors in copying or radiation bombardments of the animal's DNA which would produce two holes in the bones of the animal's head. The holes should be spherical in shape, in the front of the head, and positioned with bilateral symmetry, to be most effective. What

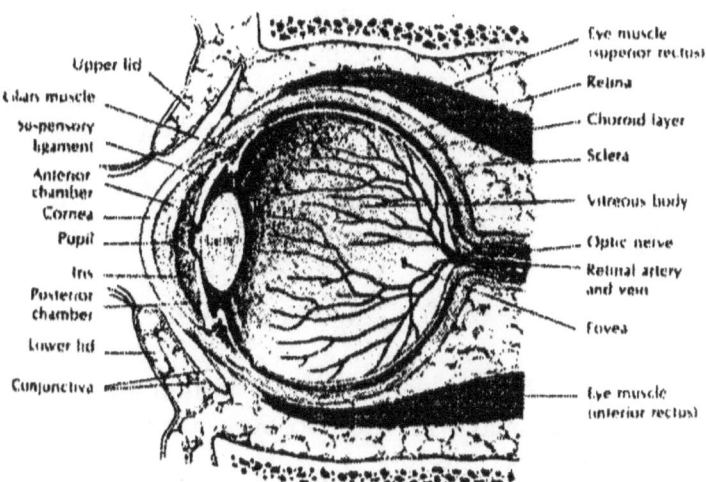

Figure 15.10 The Human Eye

161

statistical probability do you think there would be that this would take place? Although I really think that this probability would be less than one chance in a billion, for the purpose of this analysis, let's give it one chance in a million. Then, would two holes in the head provide any survival advantage? Certainly not! Natural-selection would not favor any animal with two such holes in its head. But let's proceed with our analysis anyway.

Next, two eyeballs of some description would have to suddenly appear in the two bone sockets. In the finished product of vertebrates, each eyeball is spherical in shape and consists of a triple-layer outer skin filled with a clear fluid, through which light can pass freely. The outer layer (sclera) is a tough white skin that provides protection and support. The middle layer (choroid) contains blood vessels and provides nourishment for the eye. The inner layer (retina) supports the light-sensitive rods and cones which pick up and transmit optical images. None of these features could be produced by errors in replication of the DNA or bombardment by radiation. These are *new* features that could only arise by having thousands of new atoms of carbon, hydrogen, oxygen, nitrogen and phos-phorus somehow float in, or otherwise be transported, from the surrounding land, sea, and air, and take their places as part of the DNA molecules of the animal.

Actually, we can estimate the number of new atoms that would need to be added to a gamete to produce two new eyeballs. The eyeball of a human consists of a sphere which is about one inch in diameter. Since such body parts have a density about the same as water, 0.0362 pounds per cubic inch, the weight of two eyeballs would be,

$$W = 2(4/3)(3.1416)(0.5)^3(0.0362) = 0.0379 \text{ pounds.}$$

If the human weighed 150 pounds, the fraction of the weight that is represented by the eyeballs would be,

$$F = (0.0379) / (150) = 0.000253.$$

Since a human DNA molecule has 204 billion atoms, the number of DNA atoms associated with the two eyes would be, on the really untra-conservative basis of weight,

$$N = (0.000253)(204,000,000,000) = 52,000,000 \text{ atoms.}$$

Thus, just for two eyes, 52 million atoms of carbon, hydrogen, oxygen, nitrogen, and phosphorus would need to fly away from the land, sea, and air of the earth, and take their special positions as nucleotides in the DNA of the human being. If the human being were an adult, and if the new atoms could not find a sperm cell, it would take 52,000,000 atoms for each cell, and the number of cells in an adult is about 40 trillion. (Hickman 4) What do you think is the statistical probability that these new atoms would, by chance alone, fly in from the environment and add themselves in meaningful places to the DNA molecules of a vertebrate animal?

As you consider this, please look again at the drawing of the automobile crankshaft which is Figure 15.1. Suppose you placed a blank sheet of paper on a table and then put some black ink into an empty finger-pump-type spray bottle, and then sprayed some ink into the air over the blank paper. What statistical probability do you think there would be that this process of chance would produce the drawing of Figure 15.1? I think there is about the same probability that, by chance alone, 52 million atoms from the surrounding land, sea, and air, would soar through space and add themselves to the DNA of each cell of an animal, to produce two eyeballs in the holes of the animal's head. Can you really explain how this could happen?

Consider further the following miracles of chance. In the eye of Figure 15.10, the outer tough white layer (sclera) becomes transparent in its frontal region. Consider also that there is a flexible transparent lens in the front of the eye, controlled by nerves and muscles, which can actually focus light like the lens of a camera. Consider that there is an iris and a pupil in front of the lens, controlled by nerves and muscles, which actually adjusts the amount of light that enters the eye, automatically. Consider that at the rear part of the inside of the eye of a human, there are about 125 million rods and 7 million cones which transform colored light images into stimulants which send messages concerning what is being seen along neurons to the brain. The eyeball also has nerves and muscles which enable it to rotate about any axis to aim at any object which is of interest to be seen. The eye also has eyelids which, by instinct, close, with great rapidity, to protect the eye from damage due to any object moving toward it. The eye also has its own system of lubrication.

If an animal had nothing but two holes in its head, for eye-sockets, it would not be favored by natural-selection because the holes would offer no survival advantage. If an eyeball without a lens or retina somehow came into existence in an eye-socket, it would not survive under natural-selection, because it would be useless. I have described about 15 major components of the vertebrate eye, which contain literally millions of individual minor parts. If any one of the major components were missing, the eye would be worthless, and it would be eliminated by natural-selection. Nevertheless, according to evolutionists, this eye came into existence a little bit at a time, by multiple related mutations. Certainly no one could believe that this ultra-complex perfectly-developed instrument could suddenly appear from nowhere by one mutation. That would be axiomatically preposterous. However, isn't it equally incredible to believe that there would occur a multiple series of possibly thousands of minor mutations, which would produce this remarkable instrument, the vertebrate eye. And it must be appreciated, of course, that after each mutation, the unfinished and useless eye structure would have to survive the destroying effects of natural-selection.

The major problem with multiple related mutations can be shown by the following mathematical analysis. If there were one chance in a million that mutations would produce the eye sockets, and then another one chance in a million that a mutation would produce each of the 15 major components discussed above, then the statistical probability that a series of multiple related mutations could produce a finished eye would calculate to be,

one chance in $(1,000,000)^{15}$, which equals,

one chance in 10^{90}.

This is one chance in 10 followed by 89 zeros. This means it couldn't happen. This analysis, of course, is dependent upon the arbitrary assignment of one chance in a million that each major component of the eye could be produced by a mutation. If this estimate were changed to one chance in a thousand, the result would still be preposterous. Based on one chance in a thousand, it would be, one chance in

1,000,000,000,000,000,000,000,000,000,000,000,000,000,000,000.

And this analysis relates to just one of the components of an animal, the eye. A similar study could be made for each of the other major parts of an animal. And, obviously, an eye would be worthless without a brain and a nervous system to go with it.

In connection with multiple related mutations, consider further that if the theory of evolution were, indeed, factual, and if it really were the correct explanation of the origins of all animals, then it would be acting today to produce the animals of the future. There is no logical reason that it would stop being active, ever. And, if it acted by responding to multiple related mutations, you should be able to see on your body, and on the bodies of all other animals, new parts in various stages of development. There should be thousands of such partially-developed parts. If some were only 25% finished, it is possible that you couldn't determine exactly what they were going to be. But if some were 75% completed, it should be obvious what they are going to be. Maybe you are going to get a new eye on the end of your finger. This would enable you to see well into places where you could not get your head. You could hide behind a tree and watch your enemies by sticking your finger around the tree. Maybe you will get claws on your fingers. Then you could climb trees. Maybe some of the starving people in Africa will develop additional stomachs, so they could eat grass. Can you identify one such partly-developed new feature on your body? There should be dozens of them. Can you identify even one partly-developed part on the body of any animal? There should be thousands of them.

Consider, further, that in the history of the earth there should be thousands of partially-developed parts of animals which involved bones, and the evidences of these developments should be found in abundance by paleontologists as they dig up and study fossil remains. But the fact is that no such partially-completed body parts have ever been found. All fossils appear to be the bones of completed animals, all fully developed and fully functional. Thousands of indisputable examples of partially-developed fossil parts do not exist. It is more probable that not even one exists. If you cannot point to any example of a partially-developed new part on your body, or the body of any animal, and, if there are no fossil remains of partially-developed new parts, then might it not be reasonable to conclude that multiple related mutations do not now cause, and never have caused, the evolution of new features in animals?

We may conclude, then, that multiple related mutations could not possibly produce any of the complex parts or features of an animal, for the following three reasons: (1) a series of mutations that are related to each other, and which would build, step by step, a new part, are not likely to occur by chance, (2) such new features would require that millions of new atoms be added to the DNA molecules of animals, but no logical explanation has been offered as to how these atoms would move, merely by chance, and (3) natural-selection would eliminate any part that was not completed because an unfinished part would be useless and would have no survival value. We must conclude that multiple related mutations do not qualify as a valid New-Feature-Producing Agent!

On the subject of "Mutation and Selection", Schindewolf said, "Since mutations are minute transformational steps . . . a continuum of forms would have to arise . . . the concept of species would no longer be valid . . . the result would have to be a disorganized confusion of forms diverging in all directions . . . this agrees only slightly with the . . . fossil record . . . it is no longer tenable once we take the fossil material into account." (Schindewolf 2)

Conclusion: Mutations Do Not Qualify as a Valid NFPA

We can also conclude that mutations, from any source, and all types of mutations, do not qualify as New-Feature-Producing Agents! And, the concept that mutations plus natural-selection, the primary remaining theory that many evolutionists rely upon, clearly does not constitute a plausible mechanism by which evolution could operate.

CHAPTER 16. MENDELIAN GENETICS. AN NFPA?

Mendelian Genetics Widely Publicized

I recently heard a man state on a television program that the combination of Mendelian-genetics plus natural-selection finally offered a thoroughly scientific and mathematical explanation for the mechanism of biological evolution. We must admit that this sounds sophisticated, and it could be convincing to many members of the general public. It is probably also true that many high-school teachers, textbook writers, and other evolutionists believe that the above statement is true. We know that all theories of evolution consist of two parts, (1) a description of some New-Feature-Producing Agent, plus (2) the process of natural-selection. Therefore, we must conclude that this Mendelian-genetics theory, in essence, must propose that Mendelian-genetics, or its actions, constitutes a New-Feature-Producing Agent. So, let's now study the subject of genetics.

Mendel Founded the Science of Genetics

This science was founded by Gregor Mendel (1822-1884), an Austrian monk who developed a strong interest in plant-breeding and mathematical statistics. Among his studies, Mendel performed experiments with the garden pea plant, Pisum sativum. He noted that, of pea seeds and pea pods, some were smooth and others were wrinkled, and some were yellow while others were green. He also noted that some of the pea-plant flowers were purple and others were white. He performed hundreds of experiments with peas, and other plants, using self-fertilization and cross-fertilization, and he came to three general conclusions: (1) pure characteristics of parents do not blend in offspring, but they persist to produce unaltered traits in succeeding generations, (2) the results of combining parent peas follow the laws of statistics, and (3) trait factors (genes) can be dominant or recessive, and the traits of recessive genes can reappear in multi-generation offspring.

Before describing Mendel's work, consider how surprised you would be if you mixed a quart of pure black paint with a quart of pure white paint, and found that the result was two quarts of black paint.

You probably would have thought that you would get two quarts of gray paint. Then, suppose you divided the two quarts of the resulting black paint into eight half-pint cans, and mixed pairs of these cans together to produce four pints of second-generation mixtures. Suppose you found that three of the new pint-sized mixtures were jet black, and one was pure white. You would think that a magician were present. Mendel got similar results when he bred pea plants.

An Introduction to the Science of Genetics

Here is an example of one of Mendel's early experiments. He first obtained pure true-breeding purple-flowered peas and pure white-flowered plants. He crossed them and found that all of the first-generation offspring were purple. If blending had taken place the flowers would have been light purple or white-and-purple striped. But no blending took place. He then cross-bred the above offspring and found that the second-generation of flowers were almost exactly in the ratio of three purple to one white, and none were of a mixed blend. After much study he came up with the following explanation.

He assumed that each parent plant contained two trait-determining-factors, which we will call genes. We will use the letters, P, to represent a gene which would tend to produce a purple flower, and, w, for a gene that would tend to produce a white flower. We will also assume that the genes of a plant can be represented by such symbols as P/P, P/w, and w/w, and we will assume that when a sperm or egg, or the botanical equivalents, forms in a parent it takes only one of the above genes and its letter, such as a P-sperm, or a w-egg. Mendel then assumed that a gene could be dominant or recessive, and, if an offspring inherited a P and a w, it would be purple, because P was dominant over w. Thus a pure purple flower would have P/P genes, and a pure white flower would have w/w genes. If these two parents were bred, all of the offspring would have P/w genes, and they would all be purple, which is what he observed. However, if the offspring were then bred among themselves, their sperm and eggs would produce the following possible products:

POSSIBLE COMBINATIONS	GENES THAT WOULD RESULT	RESULTING COLOR
Sperm P meets egg P	P/P offspring	Purple
Sperm P meets egg w	P/w offspring	Purple
Sperm w meets egg P	P/w offspring	Purple
Sperm w meets egg w	w/w offspring	White

Thus, because P is dominant over w, the ratio of second-generation colors would be three purple for every one white, a 3:1 ratio, and there would be no blending of the colors. Please note here the similarity between the mixing of the paint and the breeding of peas.

The above-described research, and many other similar studies conducted by Mendel, provided the foundation for the science of genetics. Mendel's work provided theories and mathematical models which succeeded in explaining many observations concerning the science of heredity. Of course much has been learned about genetics between the time of Mendel and the present, but the term, "Mendelian-genetics", is still used to refer to the body of knowledge on this subject.

DNA, Chromosomes, and Genes

To understand the subject today, it is necessary to refer to such entities as DNA, chromosomes, and genes. We discussed DNA in some detail in Chapter 15. The DNA molecule, which is present in every cell of an animal, is the drawing, or blueprint, which contains the instructions which govern the structure and function of all parts of the body. The DNA molecule is also present in the cells which ultimately become sperm or eggs, and which transmit traits to an offspring. A DNA molecule is a long threadlike structure, which, in the case of a human being, may contain 204 billion atoms. Along each DNA molecule there are many proteins attached to it like beads on a string. These beads are the only things that could be seen by early researchers using an optical microscope, and they were called chromosomes, because they appeared to be "colored bodies". A chromosome is merely a segment of the DNA molecule, and a gene is just a smaller segment of a chromosome. Human beings have 46 chromosomes in each cell, the fruit fly has 8, frogs have 26, and an amoeba has 50. Our objective here is to determine whether or not the transfer of genetic information from parents to offspring could produce new features.

Could genetic-transfer be a New-Feature-Producing Agent? To make this determination we must study in more detail the process of genetic transfer.

All of the traits of an animal are coded in its DNA, especially the segments of the DNA molecule which are called chromosomes and the sub-segments called genes. The transfer of this information in multicelled bisexual animals is accomplished by having the male parent produce a sperm, and the female parent produce an egg, and then having the two combine at fertilization. The chromosomes of a parent are paired in each cell, half of them come from the parent's father and half come from the mother, and each member of a pair cover the same type of traits as the other. These are called homologous pairs. The genes on the chromosomes are also in pairs. Both genes of a pair contain instructions for the same type of trait, but their instructions may differ in minor detail. For example, one might call for a dimple in the cheek, and the other might call for no dimple, but they both call for the same basic facial geometry. Two slightly differing genes for the same trait are called alleles.

Sperm, Eggs, and Fertilization

The process of producing sperm, eggs, and offspring takes place as follows. We will use the human being as an example. First, consider a cell in the testicle of a male parent, which cell is to become a sperm. This cell contains 46 chromosomes, called the diploid number. Actually there are 23 pairs of chromosomes. Referring to Figure 16.1, we will follow the actions of just two pairs, but the same process would be followed by all 23 pairs. In Stage 1 of the Figure, the four chromosomes of two pairs are shown. Different cross-hatching is used to represent the fact that each chromosome carries different genes.

Figure 16.1 The Production of Sperm, Eggs, and Offspring

The cross-hatched bars below represent chromosomes
in the nucleus of a potential sperm cell.

STAGE 1

A Homologous Pair of Chromosomes   Another Homologous Pair of Chromosomes

| From Male Parent's Father | From Male Parent's Mother | From Male Parent's Father | From Male Parent's Mother |

STAGE 2

Each chromosome, including all genes, is duplicated by DNA replication.

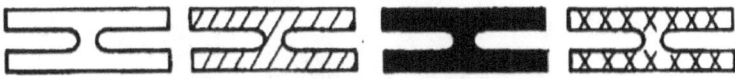

STAGE 3

Homologous
chromosomes
orient themselves
in paired alignment.

STAGE 4

Members of
each pair
exchange
genes.

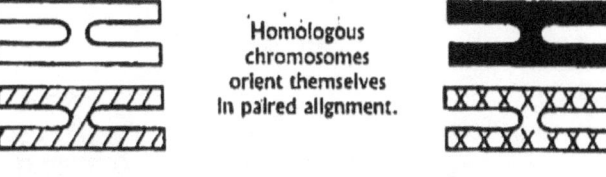

Figure 16.1 Continued:
STAGE 5

At random, individual still-attached chromatids are pulled to opposite poles of the original cell, the cell divides, and we get two cells.

STAGE 6

Flagellum
Tall

Sister chromatids become detached and all individual chromatids are pulled to a pole, and the second division takes place.

Flagellum
Tall

Flagellum
Tall

Each new nucleus, having only half the original number of chromosomes, gets a flagellum tail and becomes a sperm in search of an egg.

Flagellum
Tall

STAGE 7

Sperm   **+**   Egg   **=**   Fertilized Egg

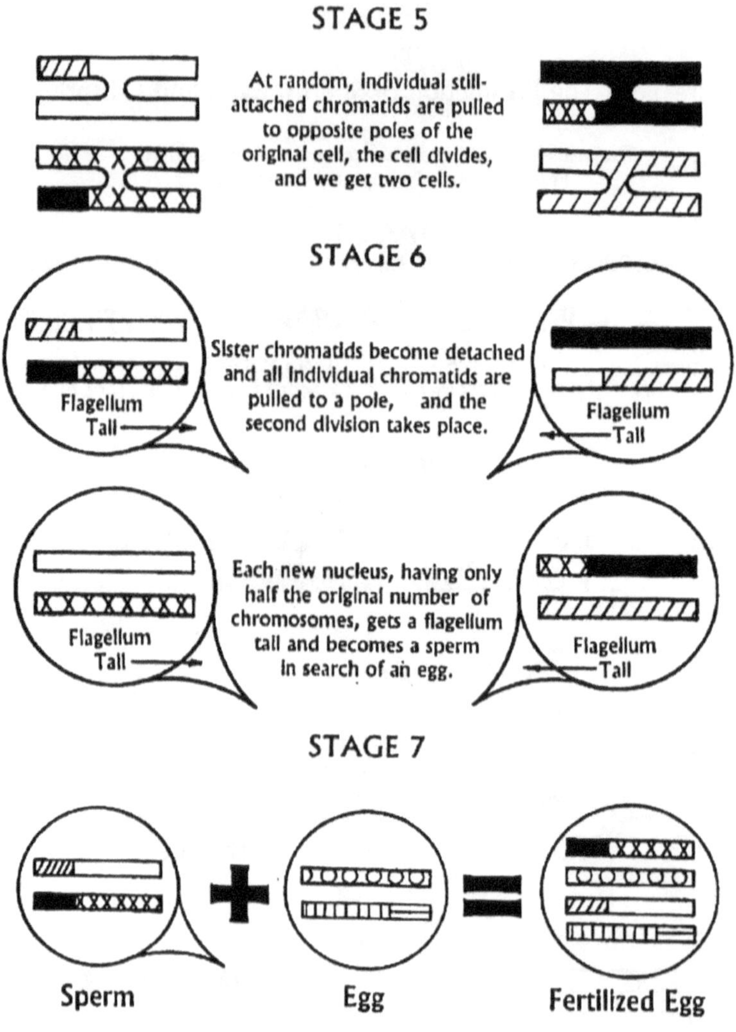

The first action that takes place in the potential germ cell, such as the sperm of a male or the egg of a female, is that the DNA of the cell is duplicated, thus producing twice as many chromosomes, but each chromosome remains attached to its duplicated sister by an attachment called a centromere. Each member of the attached chromosomes is called a chromatid. This is shown in Stage 2. At this Stage the chromosomes are randomly positioned within the nucleus of the cell. In the next action, Stage 3, all of the 46 chromosomes, with their sisters, become aligned together, with each member of a homologous pair becoming aligned next to its related member, its homologue. During Stage 4, many pairs of genes of homologous chromosomes become exchanged. Some of the genes, now close to and opposite each other, simply become detached, exchanged, and reattached. They exchange places. This is shown with the aid of the crosshatching in Stage 4. So far, all the action has taken place within the nucleus of one germ cell. Between Stages 4 and 5 the nucleus becomes polarized and microtubules develop which randomly select half the two-sister chromosomes and pull them to one pole of the nucleus, and the other half are pulled to the opposite pole. Please note that between Stage 4 and Stage 5 the homologous pairs no longer stay near each other, and the microtubules randomly pull all the still-attached chromatids to either on pole or the other. The nucleus then divides at the equator, and it separates into two new nuclei. The result is shown in Stage 5. Between Stage 5 and Stage 6 the sister chromatids finally become detached and, in each of the two cells of Stage 5, new microtubules develop and, again, they randomly pull half of the newly-formed individual chromatids, now chromosomes, to one pole, and the other half to the other pole. Divisions at the equators again take place and four new cells are the result, as shown in Stage 6. A flagellum tail then develops on each new cell and it becomes a sperm, and it swims off in search of an egg.

Incidentally, aren't you glad that your father's sperm which brought you into being had a tail on it? If it had no tail you wouldn't be here. The tail on a sperm is identical to the flagella described in Chapter 10, where we discussed protozoa. Look again at Figure 10.2 which shows the ten pairs of microtubules which act as muscles and which enable the tail to whip back and forth and propel the sperm. Do you think that these complex tails evolved? If so, how did the sperm swim before the tails evolved?

Note, now, that, whereas we started in Stage 1 of Figure 16.1 with four chromosomes, we now have four nuclei, each one of which has only two chromosomes. In our human being, whereas we started with a cell having 46 chromosomes, at this Stage we would have four germ cells, each one having only 23 chromosomes. This is called a haploid number of chromosomes, half the usual number. Note also that the chromosomes of the germ cells have had the genes and chromosomes thoroughly mixed by gene exchanges and randomly selected chromosomes in cell divisions. A similar process takes place in the sex organs of the female parent, and from each original germ cell, eggs are produced with a haploid number of chromosomes. Fertilization then takes place whereby a sperm joins with an egg, and the number of chromosomes is restored to 46, the diploid number. This fertilized cell then duplicates itself thousands of times and, directed by its own DNA, it becomes an offspring of the two parents.

Genetic Transfer can Result in Tremendous Variation

Several different conclusions can be drawn from the above description of the basic elements of animal genetics. The first conclusion is that there can be tremendous variation in the traits that an offspring can inherit from its parents. On the chromosomes of a typical animal there are thousands of genes, and millions of gene alleles. And, at Stage 4 of the process described in Figure 16.1, these genes are freely exchanged. Then, the separation of chromosomes described in Stage 5, will further mix the chromosomes as they are randomly pulled to the two separate poles of the nuclei. Then, in Stage 6, as the sister chromatids are separated, another random separation takes place. The same mixing process takes place as the female produces her egg, and, finally, of all the thousands of different sperms that are produced, only one, by chance, fertilizes the particular egg that is present in the female.

In Figure 16.2 there are shown three pairs of homologous chromosomes. This Figure shows that eight different haploid combinations can be produced from these three pairs. The mathematical representation of this is,

Possible Combinations = 2^3 = 8.

If the number of chromosome pairs were 23, as in a human, the number of possible combinations would be,

Possible Combinations = 2^{23} = 8,388,608.

Figure 16.2. Shows Eight Different Haploid Combinations of Three Pairs of Homologous Chromosomes.

The above relates only to the mixing of whole chromosomes. A similar analysis, involving the thousands of genes that could mix, would show that the possible combinations of traits would be in the billions. In fact, Francisco Ayala, a prominent biological scientist, has calculated that one pair of human beings, mathematically, could produce 10^{2017} different offspring without any two being clones. (Ayala 1) To appreciate this figure please note that there are only 10^{80} atoms in the universe. This great variation in traits, within a species, can, of course, be observed by anyone. You, the reader, have no doubt come into contact with thousands of people during your lifetime. Have you ever seen any two that looked alike, identically, other than identical twins?

Variation Applies only to Minor Traits

However, lest we become overly impressed with the wide variation that is possible within a species, we should note that this variation applies only to minor traits, not major ones. All members of a species inherit all of the major characteristics of that species. All humans have similar limbs, a face with two eyes, a mouth, and a nose. All have similar internal organs, a similar brain, similar sex organs, and similar nervous systems, circulatory systems, etc. All horses are horses, cats are cats, and pigs are pigs. By far the majority of the DNA for a given species is devoted to the major characteristics of that species, and it does not change from one individual to another. Only alleles which control minor traits are different. So, our first conclusion to be drawn from our analysis of genetics is that there can be great variation of traits within a species, but this variation applies only to minor traits.

Exchanges of Existing Genes do not Produce New Features

The second conclusion that we can draw is that the process of reproduction, governed by the laws of genetics, cannot produce new features, and hence Mendelian-genetics cannot be a New-Feature-Producing Agent. The reason for this is that within the processes of producing sperm, eggs and fertilization, the variation is only due to exchanges of *already existing* genes, the random selection of *already existing* chromosomes and chromatids, and the random success of a particular sperm in fertilizing the egg, as described above. In order to create a new feature it would be necessary to add millions of new atoms to the DNA. Exchanges and random selections do not add new atoms. Atoms of carbon, oxygen, hydrogen, nitrogen and phosphorus would somehow have to fly in from somewhere, or otherwise arrive, and take their meaningful places as new components of the DNA molecule. This is not what takes place as parents produce offspring. Review again our description of the process of reproduction. It indicates that only exchanges and random selections of *already existing* alleles, genes, chromatids and chromosomes takes place. Even some evolutionists and textbook authors agree with this conclusion. For example, after adding "mutations" to the above list of gamete-forming exchanges and selections, Cecie Starr, author of a widely-used textbook on biology, states that, "of these events, only mutation *creates* new forms of a gene. The rest shuffle *existing* genes into new combinations in new individuals." (Starr 2) But we know from our studies of mutations in Chapter 15 that they don't create viable new features either.

Variation within a Species does not Produce a New Species

Observing the tremendous variation in traits that can be introduced by sexual reproduction, governed by genetics, some evolutionists have been sorely tempted to assume that this variation has been an agent that has created new features and new species. And evolutionists have attempted to cite illustrations of this process involving peppered moths, finches, fruit flies, bacteria, skin colors, etc. However, with the knowledge available today on the subject of DNA, genes, and chromosomes, and with our modern understanding of the mechanisms of meiosis and gamete production, it is not possible to explain how this variation could produce a new trait or a new species. Furthermore, even if a new body part were created by genetic variation, it would have to be developed a little bit at a time, and natural-selection clearly would eliminate any partially-developed feature or part. And, finally, if this variation were responsible for introducing half-finished new parts, there should be thousands of examples of such developments on the bodies of the animals we see today, including humans. To my knowledge not one clear example of any such creation has been cited by evolutionists involving any animal of today, or any fossil remains of any animal of the past.

Mendelian Genetics is not a New-Feature-Producing Agent

The crucially significant limiting characteristic of genetic variation as it relates to alleged evolution, is the fact that such genetically-caused deviations are minor in nature and they result only from exchanges in already-existing genes, random polarizing selections of already existing chromosomes and chromatids, and the random success of a particular sperm finding a particular egg. It is nothing but the reshuffling of existing genes. It can cause minor variations, but it cannot produce new features which require substantial replacements or additions to the atoms in the DNA molecules of an animal.

We can conclude from this study of Mendel's work, and of modern genetics, that genetic inheritance, or Mendelian-genetics, is clearly not a New-Feature-Producing Agent.

The Cambrian Explosion

CHAPTER 17. PUNCTUATED EQUILIBRIUM. AN NFPA?

Other NFPA's

Other New-Feature-Producing Agents have been proposed in recent years as scientists increasingly have been coming to the conclusion that the proposed NFPA's of the past, principally mutations, could not really produce evolution. Realizing that the production of large changes (macro-evolution) by the cumulative effects of many small changes (micro-evolution), each of which would have to survive natural-selection, is not really a plausible theory, some evolutionists have sought refuge in such concepts as the following: (1) Animals must have evolved quickly, and large changes occurred suddenly; or, (2) Small groups of animals became separated from a larger main group, and, by adaptation, mutations, or genetic variation, the small groups evolved and became different, and then, later, the small groups rejoined the main group. Various names have been offered to represent these concepts, especially the term, "Punctuated Equilibrium." This name apparently refers to the fact that the fossil record reveals that most animals have existed without change for long periods of time, *equilibrium; punctuated* with the sudden appearance of new animals.

Many of the scientists who embrace these theories are paleontologists. One reason for this might be the fact that the fossil record does seem to support the idea that, typically, an animal may suddenly appear in the fossil record, then stay the same, unchanged, for millions of years, and then it becomes extinct. This suggests a punctuated appearance, followed by many years of stasis, equilibrium. Actually, to me, this process sounds more like the design and construction activities of a supernatural designer-craftsman than the results of evolution.

The main problem with punctuated equilibrium is that no credible explanation has been offered to tell us how such sudden large changes could be brought about by the natural forces of the earth, or how the animals of the small groups could evolve. We have shown in the previous Chapters that adaptation, mutations, and genetic variations cannot produce new features, and the scientists who believe in these punctuated-equilibrium

179

theories do not offer any convincing new explanations which describe just how the new features could be produced, either by large sudden changes, or by changes occurring in the separated smaller groups.

Punctuated Equilibrium would Require Additions to the DNA

Obviously, if a sudden and substantial change were to take place, such as the transformation of one animal into another of a completely new species, this would require that millions of new atoms would have to be added to the DNA of the first animal to produce the DNA of the second. And, each of these new atoms would have to take its very special place in the nucleotides and other chemical positions of the new DNA molecule. But, this theory does not offer any compelling explanation as to how this might be accomplished by the chance actions of the natural forces of the earth. The same is true of the theory involving small groups. For these animals to change, also, the DNA must be modified.

Prior to 1950, the German paleontologist, Otto Schindewolf, proposed a punctuated-equilibrium-type theory of evolution which he called, "proterogenesis." (Schindewolf 9) He noted that the Darwinian-type of evolution, gradual transition, was not supported by the fossil record. He suggested that evolution took place suddenly, without transition, in very early juvenile stages of development in animals, and he was able to cite several examples which he felt were supportive of this theory. However, nothing was known about DNA in his time, and he could not convincingly explain what caused these sudden changes.

Conclusion.

We must conclude about punctuated-equilibrium, that, even though it may be consistent with the fossil records, it is not a valid NFPA.

CHAPTER 18. NATURAL SELECTION. AN ANFPA?

If you have read the previous Chapters of this book you should, by now, know what is the role of natural-selection in the various theories of evolution. However, the term "natural-selection" is so widely mentioned, and so carelessly used, that I felt we needed a separate Chapter to make sure you aren't misled by the use of this term.

College Textbooks Say Natural Selection can Produce Evolution

Here are some quotations from college textbooks:

(1) "natural selection . . . leads to evolutionary change." (Hickman 5)

(2) The great contribution of Darwin and Wallace was that they "provided the first credible explanation for evolutionary change, the *principle of natural selection*." (Hickman 6)

(3) "Through natural selection new species originate." (Hickman 7)

(4) "Natural selection is the guiding force of evolution." (Hickman 8)

(5) "Evolution by natural selection is now a well-documented phenomenon in nature." (Starr 3)

(6) "Natural selection is the most important microevolutionary process." (Starr 4)

(7) "Organisms . . . , what could possibly account for their diversity? . . . a key explanation is called *evolution by means of natural selection*." (Starr 5)

(8) " . . . evolution by natural selection . . . is one of the most powerful ideas in all of science, and is the only theory that can seriously claim to unify biology." (Ridley 1)

(9)   " . . . natural selection . . . is . . . responsible . . . for the whole diversification of life from a simple common ancestor . . ." (Ridley 2)

High school textbooks contain similar statements. Ridley's textbook contains two Chapters on the subject of natural-selection. Chapter 5, entitled, *"The theory of natural selection"*, devotes 38 pages to this subject. Although textbook authors make statements such as the above, some of them seem to realize that natural selection, alone, does not constitute a plausible theory of evolution. Starr states, "Variation in traits is an outcome of certain events . . . Of these events, only mutation *creates* new forms of a gene. The rest shuffle *existing* genes into new combinations . . ." (Starr 6)

Natural-Selection's only Action is to Cause Extinctions

What, really, does natural-selection do? All it does is bring about extinctions. It doesn't create new features, parts, or species. Natural-selection brings about the survival of the fittest; but how does it do that? By causing the extinction of animals that are not fit. Natural selection not only eliminates the animals which are not fit to survive, but it increases the number of animals that are fit. An animal with an advantage will live longer and will produce more offspring, and that species will thrive. And, since life's resources are limited, the lesser fit animal will tend to become extinct. We must repeat again, every theory of evolution has to have two parts, (1) some explanation as to how new features are produced, (NFPA's), and (2) natural-selection. Natural-selection, itself, is not a New-Feature-Producing Agent!

Variation within Species is Often Mistaken for Evolution

What has caused some evolutionists to seem to misunderstand this very important concept? In the first place, Darwin called his ideas the theory of evolution by natural selection. Even to this day, Darwin's proposals are referred to as Darwin's theory of natural selection. Since Darwin's time, evolutionists have pointed to various examples as evidence that the theory of natural-selection, by itself, causes evolution.

The Peppered Moth

The most widely-used example involves the peppered moth, *Biston betularis*. This moth exists in two colors, a light color and a dark color. Before 1848, near Manchester, England, a light-colored lichen grew on the trunks of the trees, and predator birds could not see the light-colored moths on those trees; so light-colored moths flourished. Then the industrial revolution came, and its pollution killed the tree lichens, making the trunks dark colored. By 1898 the dark-colored moths predominated. Thus, by natural selection, by birds eating moths, the light-colored moths changed into dark-colored moths. Some call this evolution. Actually, it is merely an example showing that there is variation within a species, and that natural-selection tends to produce the extinction of the lesser-fit moths. A much more impressive evolutionary story would be some explanation as to how this moth could start life as an egg, then grow into a worm-like larva, then encase itself into a sack called a pupa, survive the winter, and then emerge with beautifully-colored wings which actually enable the adult to fly in the air. Do you really believe that all of this miracle of change could be accomplished by natural-selection? Or by any other process of evolution?

The Finches of the Galapagos

The finches of the Galapagos archipelago are often cited as examples of evolution caused by natural selection. These birds were studied by Darwin, and, more recently, by other evolutionists. Various finches have been observed to live on these separate islands, and birds with several different sizes and shapes of beaks have been studied, each type being particularly suited for eating a particular type of food, such as wood-boring insects, cactus flowers, fruits, and cactus seeds. The populations of the birds with specific types of beaks seemed to rise or fall depending upon which type of food was abundant in different years. Evolutionists cite this as an example of evolution by natural selection. Others contend that this is simply an example of variation within a species. To me, it would be much more impressive if an evolutionist were able to explain how a bird could evolve from a lizard.

Isolation may Modify a Gene Pool

There are many other examples of natural-selection at work, many involving species which, for some reason, became geographically isolated. Such isolated animals will tend to develop a different gene pool from which to draw their inheritance, and following the science of genetics, they may develop slightly different average traits as compared with the group from which they became isolated. The skin coloration of human beings is probably a good example of this phenomenon. In Africa, people with dark skin became dominant; but in Sweden, light-colored skin predominates. But the two are of the same species. They can interbreed and their offspring will be fertile.

Variation within Kind is Often Extrapolated

These, and other similar examples in the animal kingdom, merely demonstrate that gene pools are very large and tremendous variation is possible within a given species. But variation within a species, or geographic isolation, does not produce new features, new parts, or new, more complex, species. Evolutionists seem to eagerly embrace any concept that even remotely suggests evolutionary change, and then they grossly extrapolate this concept, and imply that such changes can explain how a simple animal could evolve into a more complex one. But, of all theories, natural-selection alone, cannot be defended as a plausible theory of evolution.

Conclusion: Natural Selection does not Create New Features

The only way new features or new species could be introduced is by the adding of millions of new atoms to the DNA of the gamete of an existing species. Can you explain how natural selection could do that? Paleontologist Schindewolf, on the subject of natural selection, said, "Selection by itself absolutely cannot create something new directly, but can only shape and develop what is already in existence . . . selectionism cannot explain why anything arises . . . Selection is only a negative principle, an eliminator, and as such is trivial." (Schindewolf 3)

Natural-selection does, indeed, take place. But all it does is cause the extinction of animals that are not fit. It then allows other animals to survive. But, obviously, it has nothing to do with creating the features that enable the more fit animals to survive. Natural-selection, clearly, is not a New-Feature-Producing Agent!

CHAPTER 19. CONCLUSIONS CONCERNING NFPA'S

Remember the Fire-ring and the Protozoa

Let's review the several alleged New-Feature-Producing Agents we have discussed. In Chapter 9 we showed that the natural forces of the earth, such as wind, rain, hail earthquake, radiation from the sun, etc., certainly could move a 10-inch rock along the surface of the earth a few feet, but these forces could not design and arrange a circle of ten such rocks to form a useful fire-ring on the top of a mountain. In Chapter 10 we saw that, millions of years ago, lightning certainly could have struck some sea-water somewhere on the earth, and some chemicals similar to organic molecules could have been produced, but there is no chance that this action could have designed and constructed a living protozoan, complete with DNA.

Mother-Nature and Teleology

In Chapter 13 we showed that the theory of evolution needs a valid and credible New-Feature-Producing Agent (NFPA) to go with the proven capabilities of natural-selection, and we studied several proposed NFPA's. We asserted that Mother Nature was nothing but a fictitious personage who is superficially credited with somehow being responsible for certain acts of supernatural creation. This credit is given by fence-straddlers who are unwilling forthrightly to admit to the existence of a real personal supernatural designer-craftsman. We also discussed Teleology and stated that it would appear to be a philosophical doctrine which is, in essence, little more than a philosopher's Mother Nature.

Adaptations

In Chapter 14 we affirmed that "Adaptation" was a term widely used by evolutionists to infer that an animal's environment has the ability to bring about new features. When analyzed more carefully, it became apparent that Adaptation was nothing other than another name for the process of natural-selection, and it has no capability whatever to design or initiate the creation of new features or new creatures. If the

185

environment could create new animals, then, in any given environment, all of the animals would be the same. We concluded that Adaptation is clearly not a New-Feature-Producing Agent.

Mutations

Then, in Chapter 15, we discussed mutations, the principal dogma upon which evolution has depended for the past 50 years. In order to understand mutations it was necessary, first, to explain the DNA molecule, the complex assemblage of billions of atoms which constitutes the plans and drawings, and the management entity, which controls the construction, growth, and functioning of each individual animal. Just as it is necessary to have engineers make hundreds of drawings which are the plans for the construction of something like an automobile, the DNA molecule, which may contain billions of atoms, constitutes the plans for the construction of an animal. If an automobile is to be changed, its drawings must be revised or augmented. If an animal is to change, its DNA must be revised or augmented. Mutations are changes brought about in a DNA molecule, usually by replication errors or by bombardment by radiation. We showed that, although mutations do occur, and they do cause changes, unfortunately, mutations could not possibly be the NFPA's which account for macro-evolution, because: (1) they are extremely rare, (2) they are almost always harmful, rather than beneficial, (3) to be beneficial, millions of atoms would, somehow, have to float through the air, or be otherwise transported, from the earth's surface, and become deposited at specific utilitarian locations in the DNA molecules, and this is inconceivable, and (4) the statistical probability of the occurrence of rare, beneficial, multiple, related, mutations, which would require the adding of atoms to the DNA, and which would require that partially-finished parts would survive natural-selection, is, essentially, zero.

Mendelian Genetics

In Chapter 16 we discussed genetics, including Mendelian genetics. We introduced the subjects of mathematical and molecular genetics. We showed that the process of meiosis, involving sperm, eggs, fertilization, and inheritance, can introduce tremendous variation in offspring, but it cannot introduce new features or new body parts that are not already in the gene pool of a species. Hence, Mendelian genetics clearly cannot be a New-Feature-Producing Agent.

Punctuated Equilibrium

In Chapter 17 we discussed Punctuated Equilibrium, and similar theories, which basically assert either that: (1) each new species evolved quickly and suddenly appeared on earth, then remained unchanged for a long time, and then became extinct; or (2) new species evolved slowly in small isolated groups, they then rejoined the main group as new species, and they then became fossilized and "suddenly appeared" in the fossil record. The punctuation refers to sudden appearance, and the equilibrium refers to the fact that new species remained in stasis for millions of years, unchanged. These theories are favored by some paleontologists because the fossil records seem to support them. However, such theories offer no explanation as to how a new species could suddenly become created, and hence these theories offer no help in our search for a valid New-Feature-Producing Agent.

Natural Selection

Finally, in Chapter 18 we explored the possibility that natural-selection, alone, could be a New-Feature-Producing Agent. We quoted in this Chapter several statements found in college textbooks which seem to assert that evolution can be brought about by natural-selection, alone. Darwin seemed to believe this. But we showed that any valid theory of evolution needs two parts, (1) an NFPA, plus, (2) natural-selection. Natural-selection does, indeed, take place. It causes the extinction of the lesser-fit animals. But it needs new animals upon which to act. Natural-selection, itself, has no capability to create new features, parts, or species. Hence, natural-selection is not a New-Feature-Producing Agent.

Conclusion: New-Feature-Producing Agents do not Exist!

So, what is the conclusion from this study of the possible existence of New-Feature-Producing Agents? It is that NFPA's capable of explaining evolution just simply do not exist! This is such a significant conclusion that it bears repeating! Do not underestimate the momentousness of this conclusion! New-Feature-Producing Agents capable of producing significant beneficial new features in animals do not exist! What this means, of course, is that there is no logical scientific basis for the theory of evolution!

Some Evolutionists Agree

Do any evolutionists recognize that NFPA's do not exist? The answer is yes. Steven Gould, the highly-respected paleontologist from Harvard, Colin Patterson, a paleontologist at the British Museum, Pierre Gross, a prominent French zoologist, Garrett Hardin, a noted American biologist, and Francisco Ayala, prominent evolutionist and author have all made statements which question the validity of the concept that evolution could occur through the cumulative effects of many small mutations. For the sake of brevity, I will quote just one of these respected scientists. Dr Gould was discussing orthodox neo-Darwinian extrapolations. This is the concept that the effects of many minor mutations could be extrapolated to produce large changes in animals, and produce new species. Dr. Gould said, "that theory as a general proposition, is effectively dead, despite its persistence as textbook orthodoxy." (Gould 1)

We Now Know Who Did Not Cause the Cambrian Explosion

Let me remind you again that the purpose of this book is to identify who made the animals. Animals are machines. I am an engineer who specializes in the design of machines. In a manner consistent with being an expert witness in court trials, I have accepted the assignment of studying the Cambrian Explosion, the sudden appearance of animals on earth. My expert-witness assignment is to find out who caused the Cambrian Explosion? Who made the animals? The past few Chapters, specifically Chapters 13, 14, 15, 16, 17 and 18, have revealed to us who did *not* make the animals, and this is part of our search to determine who *did* make the animals. Hopefully, in a later Chapter, we will, indeed, identify who caused the Cambrian Explosion!

CHAPTER 20. WHAT SCIENTISTS SAY ABOUT ORIGINS

Pertinent Written Documents

When a mechanical-engineering expert-witness accepts an assignment to study some unusual event, and determine how it happened, and who caused it to occur, he studies the machinery involved, but he must also study any written documents that pertain to the event. For example, when I studied the two tank explosions described in Chapter 2, I not only studied the tanks and the machinery which supported the tanks, but I also studied documents which recorded who filled the tanks, how many pounds of ammonia or propane had been added to the tanks, who serviced the safety valves mounted on the tanks, etc. Similarly, if there are any written documents which allegedly relate to who caused the Cambrian Explosion, we should study these documents. In fact, we would be derelict in our duty if we did not study such documents.

So, let's now ask the question, "Are there any documents in existence which shed light on who caused the Cambrian Explosion? Are there any documents which might reveal to us who designed and constructed the animals?" The answer to that question is "Yes". The documents which claim to shed light on this subject are the books of the Bible.

What is the Bible

For those of you who are not familiar with the Bible, let me tell you that it is an ancient written document which was composed over a period of more than 15 centuries by 44 authors. It contains 66 books. The Bible is primarily about God. It reveals the nature of God, and it tells us about the things that God has done. It features God's relationship with human beings. Its authors claim that their writings were supervised, and in some cases dictated, by God. But this present book is not a Bible study. Rather, it is a scientific and engineering treatise which attempts to determine who made the animals. But if our research suggests to us that the Bible may contain written documents which pertain to our case, then we must study them with our usual honesty and objectivity.

Credibility of Witnesses and Documents

Before we study what the scientists say about origins, or what the Bible says, let's discuss the matter of credibility. Law cases are won or lost on the credibility of witnesses and documents. When I first started doing expert-witness work, I was involved in a law case that went to trial. After my testimony, the attorney with whom I was working said to me that I was the most credible witness he had ever seen, and he wanted me to work with him on other cases. At first I didn't know what he meant, so he explained it to me and told me how important credibility is in law cases. I was just trying to tell the truth and make it clear to the jury. He called it credibility. He later gave me dozens of law cases, and I worked with him for many years.

In order to test the credibility of the Bible on the subject of our case, we will not restrict our study of the Bible to just what it says on the subject of who created the animals. We will study the entire scope of the subject of the origin of the universe and the earth, through to the completion of the its account on the creation of life on earth. We will study, first, what the scientists of the world say on the subject of the origin of the universe and the origin of the earth, and we will then compare this with what the Bible says on the same subjects. And, if the two are in agreement, then the credibility of both the scientists and the Bible should be enhanced. If you have a lot of confidence in what scientists say on these subjects, and, if the Bible is in agreement with the scientists, then your confidence in the credibility of the Bible should be increased.

Origins of the Universe and the Earth

Let's first discuss what our scientists have learned about the origin of the universe and the origin of the earth. When we come to a discussion of the role of water on the early earth, I will introduce some of my own thoughts on this subject. As you know, mechanical engineers are the people who design boilers and water-using power-plants to produce electricity. I have taught college courses on this subject, and I claim to know a little bit about the physics and engineering related to water, heat-transfer, and the related thermodynamics. I will present to you some novel and original thinking that I have developed as I have researched this subject. It relates to engineering analyses of the role of the waters of the earth. I think this analysis is the key to understanding what really happened on our early planet.

Among the sources from which much of the following scientific information has been derived are the following: geologists, Robert Dott and Roger Batten, (Dott 1); physicist, Gerald Schroeder (Schroeder 1); astrophysicist, Hugh Ross (Ross 2); geologist, Edward Keller, (Keller 1); anthropologist, Marvin Lubenow (Lubenow 1); and authors of the Encyclopedia Britannica. And, incidentally, I agree with the following information.

Time-period Zero. The Big Bang and the Early Earth

The Big Bang. Scientists have determined that, about 15 billion years ago, our universe suddenly came into existence in the form of an extremely high concentration of mass and energy at a temperature of about 10^{32} degrees Kelvin. This temperature is about 10-million-billion-billion times as hot as the temperature at the center of our sun. Matter and energy were in a state of fluid interchange. At that time, all of the mass and energy of the universe were concentrated at a spot in the universe smaller than a grain of sand. After a small fraction of a second the universe expanded to the size of a grapefruit, and its matter-energy complex was being propelled outward in all directions at a velocity approaching the order of magnitude of the speed of light. During this initial expansion, a force was applied to this potential matter of the universe, which imparted to it sufficient momentum to cause it to expand outward for the next 15 billion years. And the universe is still expanding today. This fantastic explosion at the beginning of the universe is widely referred to as the Big Bang.

The First Period of Fusion to Heavier Elements. Most of the matter which was produced by the Big Bang was hydrogen, the lightest and simplest of all elements. It consists of one proton and one electron. In order for other heavier elements to be produced, it was necessary for protons and neutrons to undergo nuclear fusion, a process in which atomic particles join together to produce heavier elements, but this can only take place at a specific temperature, about 10 million degrees Centigrade. At higher temperatures nuclear particles do not stay fused, they fly apart; and at lower temperatures the particles will not fuse. Following the laws of physics, as the particles of the embryonic universe expanded outward, their temperatures dropped. At about 3.5 minutes after the moment of the Big Bang there was a period of about 20 seconds during which the temperature was right for fusion, and during that period, some of the hydrogen was transformed into helium, and trace amounts of some other elements. Subsequently the hydrogen, helium and the traces raced outward to form our universe.

The Stars and the Second Period of Fusion. After several billions of years, during which the elements of the universe were rapidly moving ever outward, the next significant event in the history of the universe took place. The forces of gravity which cause all particles of matter to be attracted to each other, began to have sufficient effect to cause the accretion of mass particles toward numerous randomly distributed centers of accumulation throughout the universe. This produced the stars and galaxies. However, as the mass particles, primarily hydrogen and helium, came crashing in toward these centers, the absorption of their kinetic energies produced high temperatures at the centers of these newly-formed stars. The temperatures were high enough to cause and support, for the second time, nuclear fusions; and the resulting nuclear reactions built up the heavier elements of the universe. These reactions also produced more heat, and, in particular, much heat was derived from the fusion of hydrogen to produce helium. This is the primary process by which all of the elements of the universe were produced.

Then, after each star burned a few million or billion years, if its hydrogen, or other atomic fuel became exhausted, the star would cool and explode and spew its inventory of heavy elements out into space. Many of these atoms were then accreted into other stars, and the process of star-creation, burning, and fusion into heavier elements was repeated several times.

Accretion Forms the Milky Way, Our Sun, and Our Earth. Finally, one of the centers toward which some of these particles of debris from the explosions of former stars became attracted, was our galaxy, the Milky Way. And, within it, one of the solar systems to which debris particles, including much hydrogen and helium, were attracted, was our sun and its planets. As the particles sped toward the center of that which became our sun, they went into a spiral motion which produced a disk-shaped array, and out of this disk, by further accretion, our sun and its planets emerged. As the particles sped towards our earth, their kinetic energy was transformed into heat, and this, together with some nuclear reactions that were induced within the earth, produced enough heat to melt the earth. So then we had a hot molten earth composed of the elements that the earth was fortunate enough to attract from the space debris left behind from the terminal explosions of nearby stars. In fact we were fortunate in that our earth attracted a lot of water, some of all 92 elements, and, in particular, the elements, oxygen, hydrogen, carbon, nitrogen and phosphorus, the elements which are necessary for the existence of living creatures.

Differential Settlement Separated Earth's Constituents. While still molten, the various constituents of the earth differentially settled, so that the heavier components moved to the central core of the earth, and the lighter constituents settled to the radially intermediate region, called the mantle, and the lightest parts rose to the earth's crust and the atmosphere. While still molten, and also after solidification, through volcanic activity, the gaseous constituents and water, by a process called outgassing, rose to form the atmosphere. Actually, even to the present time, a portion of the earth's core and a region just below the earth's crust, are still molten. The major constituents of the early earth are shown in Table 20.1 below.

TABLE 20.1. CONSTITUENTS OF THE EARLY EARTH

REGION	PERCENT, BY VOLUME	MAJOR CONSTITUENTS
Core	16.2	Iron and nickel
Mantle	82.3	Ultramafic Rocks
Crust	1.4	Basaltic and Granitic Rocks
Atmosphere	0.1	Nitrogen and Water

Temperature of the Molten Earth. The temperature of the just-formed earth is estimated to have been about 3000 degrees C. About 1000 degrees C. of this is attributed to the absorption of the kinetic energy of the accreted particles, and 2000 degrees is attributed to heat generated by radioactivity within the earth. The temperature, 3000 degrees C., is slightly more than 5000 degrees F.

Atmosphere of the Early Earth. Let's consider now the atmosphere of the early earth. Before the solidification of the earth's crust, when the earth was molten, the differentiation process mentioned above not only caused the heavy iron to go to the center of the earth, but also the light gasses rose to the surface and out into the atmosphere. These gasses consisted primarily of water in the form of steam, and nitrogen. There were probably also small amounts of SO_2, CO_2, Ar, and traces of Ne, He, CH_4, and NH_3, but for purposes of simplicity we will assume that the atmosphere consisted of just nitrogen and water. The oxygen of the atmosphere came later, to be produced by the photosynthesis of the plants.

Since the outgassing process has continued, through volcanic eruptions, throughout the history of the earth, even to the present, the amount of water above the surface of the early earth was probably slightly less than what we have at the present. If we assume that the above-surface water was 80% of the present amount, that would be a reasonable estimate. We will assume that the nitrogen present then was enough to contribute 12 psi to the atmospheric pressure at the surface of the early earth. Today, the nitrogen and oxygen together produce a surface pressure of about 15 pounds-per-square-inch (psi).

Just after the earth accreted its elements from outer space, the earth was molten and its surface temperature was about 5000 degrees F. At that temperature, all of the water on earth would have been in the gaseous state. It would have been superheated steam. Since steam is lighter than nitrogen, all the steam would have been above the nitrogen. The following analyses are the results of my applications of the principles of mechanical engineering to the waters of the early earth.

Atmospheric Pressure Near the Earth's Surface. We can now calculate the surface pressure that would have existed if all of the water of the earth were in the atmosphere in the form of steam. Since 71% of our present earth is covered by water to a depth of 2.3 miles, and if 80% of this water covered the whole early earth, the average depth would have been,

$$D = (0.80)(2.3)(0.71) = 1.31 \text{ miles.}$$

If liquid water weighs 62.5 pounds per cubic foot, the pressure at the bottom of the steam in the atmosphere would have been,

$$p = (62.5/144)(1.31)(5280) = 3002 \text{ psi}$$

To this we must add the 12 psi which the nitrogen contributed. This gives us 3014 psi. In some of our calculations we will round this off to 3000 psi.

Nitrogen was Compressed Down to a 50-foot Height. One effect of this very high pressure would have been to compress the nitrogen of the atmosphere down to a very small volume. Based on the gas laws of physics and engineering the nitrogen would be compressed to a value of 12/3014, or about 0.40% of what it would be if there were no steam in the atmosphere above it. In terms of orders of magnitude, this means that the nitrogen in, let's say four miles of height, would have been compressed down to a height of only about 50 feet. In other words, at that time we had very little "air" in the sky; it was mostly steam, plus 50 feet of nitrogen; and, since nitrogen is heavier than steam, even at high temperatures, the nitrogen would have been below the steam at the surface of the earth.

Atmosphere was Largely Superheated Steam. Next, let's make a calculation to confirm that, even at the pressure of 3000 psi, the water in the atmosphere was in the form of superheated steam. If we consult a table of the properties of steam, found in any good mechanical-engineering handbook, we will find that the boiling temperature of water at 3000 psi is 695 degrees F. This means that the earth's surface temperature of 5000 degrees F. certainly would have boiled off all of the water on earth and it would have been in the form of superheated steam up in the atmosphere, above our 50 feet of compressed nitrogen. And the steam would have been stable in that status because it was constantly absorbing the radiant energy from the surface of the earth at 5000 degrees F. At 5000 degrees, the steam would rise, cool, fall, be reheated, and rise again. But it would always be steam.

Formation of Earth's Crust Caused Drop in Surface Temperature. Now let's consider what happened when the earth's surface solidified, and the earth's crust came into existence. When the hot earth radiated to the cold outer space, the earth cooled; and when the hot liquid earth cooled, its surface froze to solid rock. The solidification of the earth's crust no doubt caused a phenomenal drop in the temperature at the surface of the earth. The earth's crust is an excellent insulator against heat transfer by conduction. If you were to heat one end of a ten-foot-long rock to 3000 degrees F., using a welding torch, I imagine that you could still put your hand on the other end of the rock and not feel much effect of the heat at the hot end. Rocks are poor conductors of heat. When the earth began to solidify, and when the earth's crust was just ten feet thick, the temperature of the earth's surface no doubt dropped many hundreds of degrees. The thickness of the earth's crust now is about 100,000 feet. But when it was just 100 feet thick, it no doubt provided a layer of very effective resistance to the transfer of heat by conduction from within the earth.

As the Earth Cooled the Outer Steam Cooled. Let's now suppose that the temperature at the surface of the earth dropped from 5000 degrees to 500 degrees, which it certainly would do in a short period of time after the earth began to solidify. The temperature of 500 degrees has been arbitrarily selected just to illustrate the general phenomenon of the formation of rain drops. Let's consider the steam which was in the upper region of the atmosphere, farthest away from the earth, after the earth's temperature dropped to 500 degrees. When the steam that was heated by the earth rose from the bottom of the atmosphere, at the earth's surface, it was compressed to a very high pressure. When it arrived at the top of the atmosphere the pressure was very low. This tremendous drop in pressure would be accompanied by a corresponding drop in temperature, following the gas laws. As a gas expands due to reduced pressure, it cools. But before, when the rising steam was receiving radiation from the 5000 degree earth, it didn't cool very much. Now, with the temperature of the earth at only 500 degrees, the steam would cool substantially.

Additionally, the steam at the upper atmosphere would be radiating to outer space, where the effective temperature is very cold. Assuming a background radiation temperature then of 5 degrees K., which is about—450 degrees F. (450 degrees below zero), there would be substantial additional cooling of the upper steam due to the fact that it was radiating to a very cold outer space.

Outer Steam Cools Until it Forms Liquid Rain. This steam, high above the earth, being cooled by having its pressure drastically reduced, and by radiating to outer space at 450 degrees below zero, would soon cool to the point at which it would condense into liquid water and fall as rain. The rain drops would fall down through other steam which has been heated by the earth and is rising to take its place. It should also be noted that, after the rain drops first began to form in the upper atmosphere, the cooling effects of the upper steam radiating to outer space would be enhanced because the rain drops would shield the upper steam from the radiation coming up from the hot earth. The earth would no longer be heating this upper steam by direct radiation.

Total Darkness on Earth. One consequence of this process, in which a large percentage of the earth's water is falling as rain, would be that no sunlight could get through this huge cloud of liquid water, and there would be total darkness on the earth. From your own experience you know that a heavy rain storm produces darkness. Can you imagine the total darkness that would be produced if all, or even half, of the water of the oceans were up in the sky in the form of rain? There would certainly be total darkness!

With Further Cooling Some Rain Fell to Earth to Build its Ocean. Now let's consider in more detail the effects of having the earth's surface temperature decrease from 5000 degrees to its present temperature of about 70 degrees F. Table 20.2 gives the temperature at which water will boil at various atmospheric pressures. When the earth's surface was 5000 degrees, the earth was molten, and above it there was 50-feet of nitrogen and then several miles of superheated steam. The earth then was cooling rapidly because its 5000-degree surface was radiating, through invisible nitrogen and invisible superheated steam, to the below-zero temperature of outer space.

TABLE 20.2 BOILING TEMPERATURES OF WATER

Absolute Atmospheric Pressure at the Surface of the Earth, psi.	Temperature at which Water will Boil, degrees F.
15	212
50	281
100	328
300	417
500	467
1000	545
2000	635
3000	695

When the earth's temperature dropped to about 3000 degrees, the earth's crust began to form, and the rate of cooling drastically decreased, but the process of radiating through the nitrogen and steam continued. Throughout all of this period the atmospheric pressure on the earth's surface remained at about 3000 psi. That was the weight of the water in the atmosphere.

At lower temperatures, between 695 degrees and 3000 degrees, the temperature at the outer layer of the steam became low enough to condense some of the steam to liquid water, as described above. This water then fell as rain drops. However, in this temperature range, before it fell to the earth's surface, it was reheated by the hot earth, it boiled again into steam, and it then rose to the upper levels of the steam blanket. No rain drops fell to the earth. Then the atmosphere consisted of 50 feet of nitrogen, several miles of superheated steam, and a substantial upper layer of liquid water in the form of rain drops. The rain drops were cyclically condensing to rain, falling, becoming reheated to steam, and then rising again.

Finally, when the temperature of the earth's surface reached 690 degrees and its pressure was 3000 psi., you can see by Table 20.2 that 695 degrees was necessary to reboil the rain drops. And, if the temperature was only 690 degrees, the earth was not hot enough to reboil the rain and it then fell to earth to stay. It began to fill the ocean. Then, each time some rain drops fell to earth to stay, there would have been less water in the atmosphere, and the surface pressure would have been reduced.

When enough rain fell to lower the pressure to 2000 psi, then exactly two-thirds of the earth's water would have been in the atmosphere, as steam and water, and one-third would have been in the ocean. This is because the pressure of 2000 psi is two-thirds of 3000 psi. When the pressure dropped to 1000 psi, the ocean contained two-thirds of the earth's water. At 281 degrees and a pressure of 50 psi, 98% of the earth's water would have been in the ocean.

The Waters Above and Waters Below Controlled by Earth's Temperature. Thus, during the early earth, some of the water was in the ocean, and some was up in the atmosphere, much of it in the form of liquid-water rain drops. At first, on the very early earth, much of the water was cycling by falling as rain, and then rising as steam. This cycling continued as the earth cooled. But when the temperature of the earth cooled a little more, it was not able to reboil some of the water, it then fell to earth to stay, the surface pressure was reduced, and there was a new boiling temperature on the surface. Just enough water would fall to stay on the earth, until the pressure was reduced to the point that the boiling temperature was equal to the new lower temperature at the surface of the earth. Thus, the amount of the waters above was gradually reduced, controlled by the ever-decreasing temperature of the earth's surface.

Rain drops Help Clarify Water Phenomena. To illustrate these phenomena more clearly, let's consider a rain drop falling from the cold outer region of the steam blanket after one-third of the upper waters had fallen to the earth. At that time the atmospheric pressure on the earth's surface was 2000 psi and the boiling temperature of water, as per Table 20.2, was 635 degrees. Let's assume that the temperature of the earth's surface was 640 degrees. Then, since 640 is more than 635, when the rain drop got down near the earth it would be boiled into steam and it would rise again. Next, let's assume that the earth cooled to 630 degrees. Then when the next rain drop got down where the pressure was 2000 psi, since 630 is less than 635, the earth was not hot enough to boil off that drop so it fell to earth and was added to the ocean. Thus, as the earth cooled, the waters of the earth gradually fell out of the atmosphere.

Earth's Surface was Smooth and Devoid of Irregularities. There is another interesting consequence of the solidification of the crust of the early earth. After the molten earth solidified and cooled, it no doubt had a very smooth surface, and it was nearly spherical in shape. Its surface had no mountains or valleys, and it had no structures on it, such as trees or large rocks. It was a smooth sphere, devoid of surface irregularities. And when the waters from the atmosphere began to fall and fill the ocean, the water no doubt spread over the whole earth. It was all ocean, with no land present. And, of course, until a large amount of the waters from above fell to earth, there was complete darkness. It must have been a very bleak planet.

Earth Cooled Rapidly. There is scientific evidence that the early earth cooled very rapidly, as I have also concluded from my considerations of the great resistance to heat conduction that the early earth's crust must have presented. Scientists' studies have shown that the earth must be about 4.6 billion years old. The cooling of the earth's crust which allowed liquid water to fall to the earth, and which permitted aqueous chemical reactions to take place, is estimated to have occurred about 4.5 billion years ago. And sedimentary rocks, which could only be formed by the actions of water, first appeared about 3.8 billion years ago. This would have required many years of the actions of running water on earth. Dated at an age of 3.3 billion years ago, Barghoorn and Schoph, of Harvard, found fossils of fully formed single-celled animals in Africa (Barghoorn 1)

Time-period One. Light shines through the Upper Waters

The above-described process of rising steam and falling rain drops no doubt continued until the quantity of water in the falling rain was so low that it permitted some of the light of the sun to shine through. At first, of course, the light would have been dim and diffused, but there would have been light shining through the atmosphere for the first time on earth. The passage of light through the remaining rain drops would not have been sufficiently direct that optical images could be clearly seen through the rain. No being on earth then could have clearly distinguished such objects as the sun, moon, or stars, but sufficient light would have diffused through the atmosphere to provide an abundance of light on the earth. Also, due to the rotation of the earth, the difference between day and night would then have been clearly evident.

Time-period Two. The Expansion of the Nitrogen Firmament

As the earth continued to cool, more and more water came down from the upper regions of the atmosphere and stayed on the earth to form the world-wide ocean. As the water came down, the atmospheric pressure on the surface of the earth reduced. When all of the earth's water was up in the atmosphere, the surface pressure was about 3000 psi, and the boiling temperature of water was 695 degrees F. When the earth's temperature dropped to 417 degrees F., the pressure was about 300 psi. When the temperature dropped to 281 degrees, the steam tables say that the pressure was 50 psi.

What happened to the nitrogen when the surface pressure dropped from 3000 psi to 50 psi. Of course it greatly expanded, from about a height of 50 feet to a height of several miles. This means that we had a firmament, an expanse of the heavens, a heavenly vault, a sky, or "air" on the earth for the first time. We had a couple of miles of nitrogen as our firmament, with some waters still above it, in the form of rain and steam, and other waters below it in the form of our ever-expanding ocean.

Time-period Three. Dry Land and Plants

The surface of the early earth was being cooled by direct radiation to outer space, and due to convection involving the atmosphere, which then also radiated to outer space. When the rain began falling to earth in large quantities that probably cooled the earth additionally, and this may have cooled it rather quickly. Also, the heat-producing radioactivity within the earth diminished. Scientists estimate that the present radioactivity within the earth is less than one fifth what it was in the early earth.

Following the laws of thermal expansion and contraction, generally when an object cools it gets smaller in size. But the earth's crust, being a solid material, and having a different thermal coefficient of contraction as compared with the earth's interior, would be caused to shrink and wrinkle up as the earth cooled. This phenomenon might be compared with a plum changing into a prune, or a grape turning into a raisin. A British astronomer has recently estimated that the total shrinkage of the earth since its origin has been more than 200 miles in radial dimension (Dott 2). Obviously, as the earth became wrinkled, portions of the earth would have risen above the sea level, and other portions would have produced deeper waters. And so, instead of having only sea water covering the whole earth, we then had dry land and seas.

Then, with dry land on the earth, the processes which produce soil no doubt began, and once soil was produced, the plants of the earth probably began to flourish. In fact, at that point in the history of the earth, the earth had an abundance of rain, a warm climate, and little oxygen in the air. It is conceivable that the plants grew in great abundance. And, with little oxygen in the atmosphere, when the plants died, there was no oxygen to cause them to oxidize and decay and disappear. They would just accumulate on the ground. It is possible that this abundant oxygen-deprived vegetation, when covered with other earth, brought forth our resources of coal and oil.

As the plants flourished, they produced oxygen by the well known process of photosynthesis. The oxygen content of the air today is believed to have been largely the result of this process of photosynthesis provided by the plants of the earth. Time-period three produced dry land, seas, and plants.

Time-period Four. Sun, Moon and Stars Become Visible.

As the earth cooled further, it finally reached a temperature of about 200 degrees F. At that temperature, the corresponding boiling-point pressure was 12 psi. At that pressure and temperature most of the waters which had been rising as steam and falling as rain would have fallen to the earth, and the nitrogen, with whatever oxygen had been accumulated, would have filled the atmosphere, as it is today.

With most of the water down out of the atmosphere, visibility upward from earth would be very clear, and the sun, moon and stars became clearly distinguishable from the earth, as they are today. This produced a distinct difference between days and nights, and the movements of the moon, stars and planets later gave inhabitants of the earth celestial bodies which could be used to demarcate the seasons of the year.

Also, when most of the waters came down out of the sky, the earth changed from being a planet dominated by the heat produced from the earth itself, to a planet on which the heat was dominated by the radiation from the sun. This meant that we began to have seasons, summer, fall, winter and spring. And the average temperature at the equator became very warm, but at the poles it was very cold. The pattern of water distribution became very different. Instead of general rainfall everywhere, water was vaporized largely near the equator, and distributed in the form of clouds and rain to all other parts of the world. Wind and weather became important forces of earth.

Time-period Five. Creatures of the Seas.

Scientists generally agree that the creatures of the sea were the first animals to appear on earth. Fossils of single-celled animals were found having an estimated age of 3.3 billion years. At about 540,000,000 years ago the Cambrian Explosion occurred. Suddenly, thousands of marine invertebrates appeared. Then, between 540 mya (million years ago), and about 300 mya, numerous other marine creatures came into existence, including trilobites, jawed fishes, freshwater fish, corals, bryozoans, sharks, and amphibians. This was the era of sea creatures.

Time-period Six. Land Animals and Man.

During the period from about 300 mya to about 8,000 years ago, the land animals came into existence. This included reptiles, insects, dinosaurs, placental mammals, apes, monkeys, whales, grazing animals, large carnivores and man-like creatures, such and the Neanderthal.

Scientists place the first appearance of birds at about 150 mya, which is the age assigned to Archaeopteryx, the fossil remains of a large bird. (Hickman 9).

Time-period Seven. No New Animals Since Modern Man Appeared.

No fossils or bones of any new animals have been found since the appearance of modern man, about 8,000 years ago.

The Credibility of the Scientists

As stated before, the scientific information contained in this Chapter has been drawn from the publications of many scientists, all of whom I respect. Their conclusions are based on thousands of hours of research using instruments worth millions of dollars. In most cases independent methods have come to the same general conclusions. For example, with respect to the Big Bang, scientists have measured the velocities of many stars and their distances from us. And, by extrapolating backwards, they have found that they all came from a common point at a common time.

Independent studies of background radiation and red shift confirm these findings. Similarly, the accounts of the origin of the earth are based on much corroborative data. All of these arguments, when studied in detail, are very convincing. With respect to the origins of the universe and the earth, these scientists have no motive to deceive us.

But let me remind you again that what is said in this Chapter about the role of water in the early earth is an engineering analysis made by me, your Author. I think you will find it interesting and convincing. It must have happened just as I have related it to you. But I want you to know that this analysis is my contribution, not the conclusions of the other scientists.

CHAPTER 21. WHAT THE BIBLE SAYS ABOUT ORIGINS

In the previous Chapter we presented what the scientists of the world have to say concerning the origins of the universe, the earth, and our plants and animals. In this Chapter we will examine what is recorded in the only credible ancient written document which covers these same subjects, the Bible. We make no apology for the necessity to refer to the Bible. It is simply the only credible written document which contains information relative to our case, and we would be grossly neglectful of our duty if we ignored it. We will find the relevant documents primarily in the book of Genesis, the first book of the Bible. We will organize this material into the same time periods we used in the previous Chapter, but, in the Bible, these periods are referred to as "days".

Before the First Day.

The following is recorded in the first Chapter of the King James Version of Genesis:

1. In the beginning God created the heaven and the earth.
2. And the earth was without form, and void; and darkness was upon the face of the deep. And the Spirit of God moved upon the face of the waters.

Scientists can take us back through 15 billions of years of history, right back to the Big Bang, but they cannot tell us what brought about the Big Bang. They can tell us a lot about the nature of the Big Bang, and what happened after it occurred, but they can offer no explanation as to what caused the Big Bang. They freely admit that the Big Bang cannot be explained in terms of the natural forces of the universe. Physics and chemistry cannot explain it.

But, the Bible simply says, "In the beginning God created the heaven and the earth." That was the Big Bang! Our written document says that God did it. Since natural forces didn't do it, then it must have been done by some being. We have a legal document which says that God did it. This is legitimate testimony. If He didn't do it, then who did do it? It is credible to believe the document, and conclude that God caused the Big Bang.

Our document then says, "and the earth was without form and void." This is in complete agreement with the scientific evidence of the previous Chapter. After the molten earth first solidified and formed the earth's crust, it was smooth and spherical. The word, "form", means shape, contour, or structure. The early earth had on it no trees or rocks. It was smooth, and without form. It was devoid of irregularities. The word, void, means unoccupied, empty. These words give an excellent description of the early earth.

The document then says "and darkness was upon the face of the deep." As explained in the previous Chapter, there was darkness on the early earth because, at that time, the earth had up in the top of its atmosphere millions of raindrops. No light could get through from the sun to the earth. The term, "the face of the deep" refers to the waters of the ocean that was being accumulated as the earth was cooling and some of the upper waters were coming down to earth to stay. Notice that "the Spirit of God moved upon the face of the waters." As our scientific account of the previous Chapter relates, at that time there was no land. He had no land on which to stand. So we find God moving near the surface of the waters. God was probably moving over the waters to view what He had created.

First Day. Light Shines Through

3. And God said, Let there be light: and there was light.
4. And God saw the light, that it was good: and God divided the light from the darkness.
5. And God called the light Day and the darkness he called Night. And the evening and the morning were the first day.

Especially in view of our engineering analysis of the role of water on the early earth, these verses are truly remarkable. As explained in the previous Chapter, when the temperature of the earth cooled down enough so that some of the rain fell to earth to stay, there was less rain in the upper-waters region, and light could shine through from the sun. And, although this first light was diffused, it was sufficient to light the earth and cause a noticeable difference between day and night.

Second Day. Firmament Expands Between the Upper and Lower Waters

6. And God said, Let there be a firmament in the midst of the waters, and let it divide the waters from the waters.
7. And God made the firmament and divided the waters which were under the firmament from the waters which were above the firmament: and it was so.
8. And God called the firmament Heaven. And the evening and the morning were the second day.

These verses are truly remarkable. As explained in the previous Chapter, the very early earth had a temperature of about 5000 degrees F., and all of the waters of the earth were up in the atmosphere in the form of steam. The weight of all of this steam produced a pressure on the nitrogen below of about 3000 psi. This compressed the nitrogen down to a height of about 50 feet. But later, after much of the upper waters had fallen to the earth, after the earth cooled substantially, this pressure was greatly reduced, as from 3000 psi down to 50 psi. Then the nitrogen, which was between the upper waters and the lower waters, would have expanded, from 50 feet in height to several miles in height. This produced "a firmament in the midst of the waters", and it "divided the waters which were under the firmament (the ocean) from the waters which were above the firmament (the raindrops condensing from the upper steam)." My dictionary defines the firmament as "the expanse of the heavens." God called it "Heaven".

Third Day. Dry Land and Plants.

9. And God said, Let the waters under the heaven be gathered together unto one place, and let the dry land appear: and it was so.
10. And God called the dry land Earth; and the gathering together of the waters called he Seas: and God saw that it was good.
11. And God said, Let the earth bring forth vegetation, the herb yielding seed, and the fruit tree yielding fruit after its kind, whose seed is in itself, upon the earth: and it was so.

12. And the earth brought forth vegetation, and herb yielding seed after its kind, and the tree yielding fruit, whose seed was in itself, after its kind: and God saw that it was good.
13. And the evening and the morning were the third day.

Again, as explained in the previous Chapter, as the earth cooled, its crust wrinkled up, as would be expected. This caused some of the earth's crust to rise above the surface of the ocean. So we had dry land. Other portions of the crust deformed downward and this made deeper seas. Then we had dry land and seas. Our document, the Bible, says "let the waters under the heaven (the ocean) be gathered together unto one place, and let the dry land appear."

Then, after this dry land became fertile, under the actions of weathering forces, the next thing that might be expected would be the appearance of plants. The Bible simply says, "And God said, let the earth bring forth vegetation . . ." The Bible then describes the herbs and trees in some detail, and concludes that the third day has ended. Scientists generally agree that the appearance of plants on earth preceded the appearance of the animals. The plants were necessary to provide food for the animals, and they produced the oxygen in the air which was necessary for the survival of the animals.

Fourth day. Sun, Moon and Stars Became Visible

14. And God said, Let there be lights in the firmament of the heaven to divide the day from the night; and let them be for signs and seasons, and for days, and years;
15. And let them be for lights in the firmament of the heaven to give light upon the earth: and it was so.
16. And God made two great lights; the greater light to rule the day, and the lesser light to rule the night: he made the stars also.
17. And God set them in the firmament of the heaven to give light upon the earth,
18. And to rule over the day and over the night, and to divide the light from the darkness: and God saw that it was good.
19. And the evening and the morning were the fourth day.

From the upper waters, after enough raindrops had fallen to earth to stay, so that the rain had little diffusing effect on the light passing through it, then the optical images of the sun, moon and stars would have become clear, and these objects could be seen from the earth. Of course, God created these objects earlier, when, "In the beginning God created the heaven and the earth (the universe)." But it was not until the fourth day that the sky was clear enough for them to be visible from the earth. Recall that God was on earth "moving upon the face of the waters." On the fourth day He could clearly look up and see the sun, moon and stars. The point of view was from the earth looking upward. In these verses, the Hebrew word, "asah", which is translated, made, is really a word which connotes that it was made in the past, a completed action (Ross 3).

Fifth Day. Creatures of the Sea.

20. And God said, Let the waters bring forth abundantly the moving creature that hath life, and fowl that may fly above the earth in the open firmament of heaven.
21. And God created great sea monsters, and every living creature that moveth, which the waters brought forth abundantly, after their kind: and God saw that it was good.
22. And God blessed them, saying, Be fruitful, and multiply, and fill the waters in the seas, and let fowl multiply in the earth.
23. And the evening and the morning were the fifth day.

This Biblical account clearly states that during the fifth day God created "great sea monsters" and "every living creature that moveth, which the waters brought forth . . ." Scientists agree that the first animals that appeared on earth were sea creatures. They appeared in abundance between 540 mya and 300 mya.

However, the Bible states that birds, also, were created during this period. But scientists are of the opinion that birds came later. However, that opinion is based on fossil records, and it would not surprise me if the scientists are wrong on this particular subject because of the fact that birds don't tend to get trapped into becoming fossils. Birds can fly away from catastrophes that might entrap land or sea creatures. It is possible that birds have been in existence longer than the scientists think; they just didn't get trapped into becoming fossils.

Sixth day. Land Animals and Man.

> 24. And God said, Let the earth bring forth the living creature after its kind, cattle, and creeping thing, and beast of the earth after its kind: and it was so.
> 25. And God made the beast of the earth after its kind, and cattle after their kind, and every thing that creepeth upon the earth after its kind: and God saw that it was good.
> 26. And God said, Let us make man in our image, after our likeness; and let them have dominion over the fish of the sea, and over the fowl of the air, and over the cattle, and over all the earth, and over every creeping thing that creepeth upon the earth.
> 27. So God created man in his own image, in the image of God created he him; male and female created he them.
> 31. And God saw everything that he had made, and, behold, it was very good. And the evening and the morning were the sixth day.

First, the Bible states that on the sixth day He created "living creatures . . . cattle and creeping things, and beasts of the earth . . . and everything that creepeth upon the earth." These are obviously the land animals. As reported in the previous Chapter, scientists agree that the land animals appeared next in the record of the fossils and bones. The order of events is the same in the Bible as in the fossil records. Land animals appeared in abundance between 300 mya and 8,000 years ago.

Finally, God said, "Let us make man in our image . . . and let them have dominion over the fish . . . the fowl . . . and over every thing that creepeth on the earth . . . So God created Man . . . male and female created he them." Again, the scientists agree that modern man appeared as the last creature produced on earth. Modern man appeared about 8,000 years ago.

Seventh Day. Rest. Creative Work Ended.

The following portion of our document is from Chapter 2 of Genesis.

1. Thus the heavens and the earth were finished, and all the host of them.
2. And on the seventh day God ended his work which he had made; and he rested on the seventh day from all his work, which he had made.
3. And God blessed the seventh day, and sanctified it, because that in it he had rested from all his work which God created and made.
4. These are the generations of the heavens and of the earth when they were created, in the day that the Lord God made the earth and the heavens,
5. And every plant of the field before it was in the earth, and every herb of the field before it grew,

Our document, the Bible, here states that "the heavens and the earth were finished" and "God ended his work which he had made." He repeats that he made "every plant of the field before it was in the earth, and every herb of the field before it grew." The Bible then states that "he rested on the seventh day from all his work which he had made." He identified the seventh day as a day of rest, and we still follow this pattern, even to this day. Sunday is a day of rest.

It is significant to note that the seventh day was not demarcated with the phrase, "and the evening and the morning were the seventh day". Presumably, the seventh day has not ended. In other words, God indicated that his work of creation was finished and it would not resume. He "ended his work." And this is verified by our scientists. Since the appearance of modern man 8,000 years ago, no fossils or bones of any new creations have been found.

The Remarkable Correlation Between Science and the Bible

And now, my dear Reader, are you impressed with the fantastic correlation that exists between what scientists and engineers have learned about the universe, and what the Bible says about the universe? In my opinion the correlation is profound, even uncanny. Let's examine further the truly remarkable extent of this correlation.

Who wrote the book of Genesis? It is credited to Moses. How did he learn the things he wrote in those first two Chapters? Was he there? Did Moses witness the origin of the universe? Was he learned in science and engineering? Did he have steam tables? Did he have a Hubble telescope? Of course he did not have any of these! But yet the account that he wrote contains the same items and is in the same order as the findings of our most modern scientists. Moses wrote Genesis more than 3000 years ago. Our scientists discovered the information summarized in the previous Chapter largely within the past 100 years. How could Moses have learned about the origins of the universe, the earth and life on earth?

Let's make a list of the significant entities that Moses identified in the first two Chapters of Genesis.

1. The <u>creation of the universe.</u> The Big Bang.
2. The earth was without <u>form.</u>
3. The earth was <u>void.</u>
4. There was <u>darkness</u> on the early earth.
5. God moved on the <u>face of the waters.</u>
6. Diffused <u>light</u> then shone through.
7. <u>Day and night</u> came into existence.
8. A <u>firmament</u> developed.
9. The firmament had <u>waters above and waters below.</u>
10. <u>Dry land</u> appeared.
11. <u>Plants</u> appeared.
12. The <u>sun, moon and stars</u> became visible.
13. The stars could be used for <u>signs and seasons.</u>
14. <u>Sea creatures</u> appeared.
15. <u>Land animals</u> appeared.
16. <u>Modern man</u> appeared.
17. <u>Rest.</u> Creation stopped. No more creatures appeared.

It should be appreciated that, in order for Moses to describe the items in the above list, he not only had to learn what each item was, but he also had to learn the order in which these events occurred. To fully appreciate what Moses did, let's make a couple of statistical analyses. First, let's calculate the statistical probability that Moses could have guessed the identities of the items on the above list. It is conceivable that Moses could have guessed what several of the items on the list might have been; but there are many of them that we know he could not have identified with the knowledge he had. Let's assume that Moses could have guessed that the following items should be on the list: the creation of the universe, and the appearances of plants, sea creatures, land animals and modern man. This leaves 12 items that he would have had great difficulty identifying. He certainly had no knowledge of the temperatures or pressures on the early earth, the role of water in the earth's history or the lighting conditions on the early earth. Let's assume further that there would be an average of one chance in 10 that he could guess what any of these 12 items was. I think this is a very conservative assumption. Then, the statistical probability that he could guess the 12 remaining items would be one chance in 10^{12}. This amounts to

One Chance in 1,000,000,000,000.

Next, let's estimate the statistical probability that Moses could have guessed the proper order in which the events occurred. Let's assume that he could have guessed that the creation of the universe was the first item. But Moses really knew nothing of the other events. He wasn't even born until the time of the second Chapter of Exodus, and he certainly couldn't determine the order of these events by reasoning. So then, statistically, there would be one chance in 16 that he could have guessed which one of the events came second. Then there would be one chance in 15 that he could have guessed which was third; and one chance in 14 that he could have guessed the fourth event. Following this logic, we can calculate what would be the statistical probability that Moses could have guessed the correct order for all 17 events. It would be one chance in,

(16)(15)(14)(13)(12)(11)(10)(9)(8)(7)(6)(5)(4)(3)(2)(1).

This calculates to

One Chance in 21,000,000,000,000.

From these statistical analyses we are forced to conclude that Moses, himself, did not conceive nor generate the information contained in the first two Chapters of Genesis. But the information is there. You can read it for yourself. It is an accurate and concise account of the origins of the universe, the earth and living plants and animals. And it is in agreement with what our scientists have discovered during the past 3000 years.

Elsewhere in this ancient written document we call the Bible, it says,

"Holy men of God spoke as they were moved by the Holy Spirit (God)." In other words, God told Moses what to write. In hundreds of other places in the Bible it is claimed that God, the Being who created the universe, dictated or supervised the writings of the authors of the document. It would appear, then, that we must conclude that God wrote the first two Chapters of Genesis, and Moses was just His secretary.

If we study carefully and objectively the correlation between the accounts of the scientists and the Bible, and if we take into consideration the statistical analyses presented above, we must certainly conclude that this ancient written document is a very credible one, indeed. And, if we are forced to believe what it says on the subject of the universe and the earth, then we should also take seriously what it says about who caused the Cambrian Explosion.

CHAPTER 22. WHAT THE FOSSILS SAY ABOUT ORIGINS.

Dinosaurs.

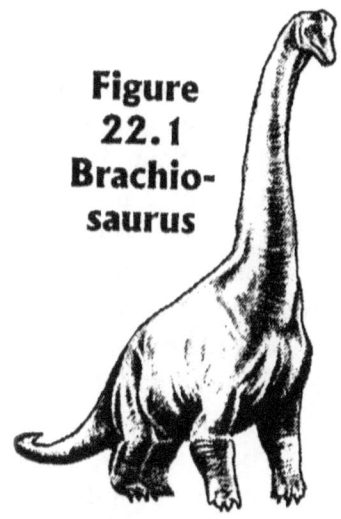

Figure 22.1 Brachiosaurus

In 1677 a huge bone, partially buried, was found in England. It was a fossilized bone from the dinosaur, Megalosaurus. Before that discovery, no one knew that dinosaurs existed. Since then, thousands of dinosaur fossils have been found all over the world. More than 800 species of dinosaurs have been identified. According to scientists, the dinosaurs appeared in the fossil record about 200 million-years-ago (mya), and they suddenly disappeared in the mass extinction of 65 mya. Some dinosaurs were herbivorous, plant-eating animals, including Brachiosaurus, shown in Figure 22.1, (Craig 3) which was 42-feet high, 91-feet long and weighed 110 tons. Others were carnivorous predators, including Tyrannosaurus Rex, shown in Figure 22.2, (Craig 4) which was 18-feet high, 39-feet long, and weighed 6.5 tons. These were powerful and ferocious animals and they soon dominated the food supplies of the earth.

Then, about 65 mya, it is theorized by one group of scientists, including L. W. Alvarez (Alvarez 1), that a huge asteroid several miles in diameter, struck the earth. This produced a huge cloud of dust, including particles from the earth and from the asteroid, which shielded the earth from the sun's rays for many months. This killed most of the plants on land and the marine algae of the seas, all of which depend on the sun's rays and photosynthesis for survival. The resulting lack of plant food killed most of the land and sea creatures of the earth. The fossil record indicates that all of the dinosaurs and most of the other land and marine animals were killed about 65 mya. Further evidence of this theory and of this mass extinction is based on the element, Iridium. The earth has very little of this element, but many asteroids are known to contain a much greater percentage of this substance. Throughout the world, scientists have found a thin layer of Iridium-rich rock, and the dating of this layer is exactly at 65 mya, just when the dinosaurs became extinct.

Figure 22.2 Tyranno-saurus Rex

It is from fossils that we can learn about the animals which once lived on earth, and which have become extinct. It has been estimated that more than 95% of all the animal species which have ever lived on earth have become extinct (Craig 1) (Godfrey 1); so that the species living today represent less than 5% of all the species which have ever existed. Scientists can add their knowledge of existing animals to that which comes from fossils and produce much information about the prehistoric animals. But this book seeks, primarily, to gather information about the origins of the animals, not their biological characteristics. So we must ask, "What do the fossils say about the origins of animals?"

If you have read Chapters 13 through 19 of this book, or if you have otherwise objectively studied the various mechanisms that evolutionists have proposed to explain how new features could be produced in animals, you should have come to realize that there does not exist any credible theory to explain how new features could come into existence, or how animals could have evolved. But then you might ask, "Maybe I don't understand how animals could evolve, but isn't it true that the fossils prove that animals did, in fact, evolve?" This is the question we seek to answer in this Chapter.

What Are Fossils?

Animal fossils are the remains of ancient animals preserved in the earth's crust. Some examples of fossils are shown in Figure 22.3. In most cases, a fossil results if an animal is suddenly buried, either on land or at the bottom of the sea, and then, with the passage of time, the soft parts of the animal decay and are lost, but the hard parts, such as bones, teeth, and shells, persist because dissolved minerals of the earth replace the cells of these parts, and their shapes are retained to form rock-like replicas. In some cases soft parts of buried animals can form a casting mold which eventually may leave an imprint on a rock. These fossils are found by digging in the earth, and, now, after more than a century of extensive research and recovery, "There are a hundred million fossils . . . in museums around the world." (Kier 1) "About 250,000 fossil species have been identified." (Godfrey 2)

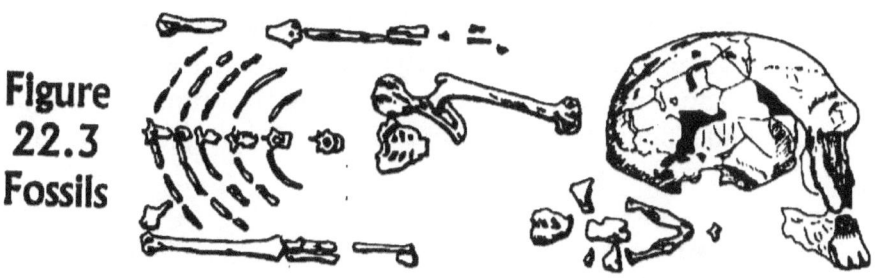

Figure 22.3 Fossils

Fossils are found only in sedimentary rocks, such as sandstones formed from sand, shales from silt or clay, or limestones from the shells of marine organisms. We do not see fossils forming today. Most fossils were probably formed when some catastrophe occurred which was capable of suddenly burying the animals and then keeping them buried for a long period of time. Scientists state that rocks of 3.5 billion-years-ago (bya), contain fossils of well-developed single-cell animals. Then many fossils of marine invertebrates appeared about 540 mya, followed by fossils of all phyla of the animal kingdom spread over the years until the recent past.

The Dating of Fossils.

Although the primary objective of this book is to determine who made the animals, not when they were made, nevertheless, when speaking of fossils it is convenient to have some estimate of when the animals lived that produced the fossils. The methods used by scientists to determine the ages of rocks and fossils include: (1) the use of index fossils and the geologic column, and (2) the use of the rates of radioactive decay of certain elements. There is no question but that either of these methods, when used on any specific rock or fossil, could produce an erroneous result; but, in scientific or engineering work, even if there is a lot of scatter in the plotting of some data, and some points might be far off of the average curve, if there is a clear trend, and especially if it is supported by more than one independent data source, the resulting plot is probably reasonably accurate. In the case of the dating of rocks and fossils, millions of age determinations have been made, and, in spite of the fact that there may have been many errors in individual cases, I think it is reasonable to rely on the ages and dates established by the scientific community. Therefore we will assume that the earth is about 4.6 billion years old, the animals have existed for about 540 million years, the dinosaurs became extinct about 65 mya, hominids have existed for 4 million years, etc. Anyone interested in the subject of timing and dating should read the book, *Creation and Time,* by astrophysicist, Hugh Ross.

Theories of Animal Origins.

In order to study the fossils and determine what they say about origins, we will need to refer from time to time to the two theories of origins. Therefore we must very carefully define these two theories. Then we will consider, one at a time, the various characteristics of the fossils, and we will test each of the two theories to see which one can best account for that characteristic. I should probably warn the Reader again, here, that in this Chapter, and in the following Chapters, I may from time to time need to refer to God. May I again plead with you to understand that this is not a religious book, based on faith and doctrinal beliefs. It is an engineer's attempt to get at the truth concerning who created the animals. But, as I clearly stated on the first page of Chapter 20, an expert witness is absolutely obligated to study any written documents which pertain to the law case he is investigating; and, in this case, such documents include the books of the Bible. And the Bible speaks of God. So I must write about God. As an engineering expert, if I discover that God and the Bible are involved in my law case, I must deal with them.

The Theory of Evolution.

For this study we will use the theory of evolution that is most widely proclaimed by the scientists and teachers of the world. This theory asserts that every animal species evolved from some similar, but less complex, ancestor by a gradual process consisting of many small changes brought about by some new-feature-producing agent, such as mutations or genetic variations, followed by the pruning action of natural selection, but where all of the above-identified processes take place by the chance actions of the natural forces of the earth without any help from any being.

The Theory of Creation.

The theory of creation to be used in this book asserts that every animal species of the earth was designed and constructed by the supernatural being, God, specifically the God of the Bible, with no evolutionary forces of any kind involved, and that God came down to the earth from time to time to design and construct plants and animals, during the past several million years, after the earth cooled sufficiently to support life.

Both Theories Relate to Whether or Not God Exists

One of the primary concepts of the theory of evolution is the assumption that God does not exist; and one of the primary concepts of the theory of creation is that God does exist. Although the matter of God's existence will be mentioned again in Chapter 24, since it is crucial to the subject of fossils, we will also address it briefly here.

As explained in Chapter 20, the universe came into existence with the Big Bang about 15 billion years ago. According to the theory of creation used in this book, God caused the Big Bang, and he then supervised the development of the universe through the following stages:

(1) The accretion of elements to produce stars and galaxies.

(2) The nuclear fusions in the stars by which the heavier elements were produced.

(3) The explosions of the older stars which produced matter debris in the universe.

(4) The accretion of the multi-element star debris which produced the earth.

(5) The design and construction of the plants and animals on the earth.

In particular, I believe that after the earth cooled enough to permit life, God, from time to time, came down to the earth to design and construct the animals. What evidence is there to support this contention? First, it is consistent with the record of the fossils. Second, it is consistent with God's behavior as recorded in the Bible. The Bible is an ancient written document which is at least as reliable as the testimony from any other ancient writing. The following quotations are from the Bible.

> "And the Spirit of God moved upon the face of the waters." (Genesis 1:2)
>
> "And they (Adam and Eve) heard the voice of the Lord God walking in the garden . . ." (Genesis 3:8)
>
> "And the Lord came down to see the City . . ." (Genesis 11.5)
>
> "And the Lord appeared unto him, (Isaac) . . ." (Genesis 26:2)
>
> "And the Lord went up . . ." (Genesis 35:13)
>
> "the children of Israel sighed . . . and their cry came up unto God . . ." (Exodus 2:23)
>
> "on the third day . . . the Lord will come down upon Mt. Sinai . . ." (Exodus 19:20)
>
> "And the Lord came down in a cloud . . ." (Numbers 11:25)

Apparently, *after* the creation of man, it was a common practice for God, from time to time, to come down to the earth or to go up from the earth. He obviously had interest in the earth and its inhabitants, and it would be logical to assume that he probably also came to and from the earth during the time period *before* man was created, during the period that all of the animals were being created. Based on God's pattern of visits to the earth, we can speculate that, from time to time, he came down to the earth to create new animals, especially during the period from 540,000,000 years ago to 8000 years ago. Therefore, in this book, when we speak of the theory of creation, we intend that this term should include all of the acts of God described above.

Credibility of Theories to be Tested by Considering Fossil Characteristics.

The purpose of this Chapter is to study the extent to which the fossil record supports either the theory of evolution or the theory of creation. In each of the Sections below, a specific characteristic of the fossils will be identified, and the extent to which that characteristic supports one or the other of the theories will be explored. Each of the following Sections will cover one fossil characteristic.

Fossils can be Arranged Into a Family Tree.

Scientists have organized the animal kingdom into the following categories: phylum, class, order, family, genus and species. If all of the species of the animal kingdom were plotted on a large map, the result could be interpreted as a family tree, because it would suggest relationships among the various members of the tree. Figure 22.4 illustrates the concept. (Snelling 1) The cataloguing would be based on similarity of animal characteristics. The fossils of animals can be catalogued into the same categories as have been defined for animals living today.

Figure 22.4 Family Tree

Evolutionists claim that the fact that all of these past and present animals can be organized into a tree is an argument suggesting that they evolved one from another. Creationists suggest that the fact that some animals are similar to others, and some animals lived earlier than others, and that they can be organized into a tree, merely reflects the design thought processes of God as he leisurely designed and constructed the animals during his visits to earth over the past several millions of years.

We could organize a series of tacks, pins, nails, screws and bolts into a tree which would show the evolution of fasteners, but all it would really indicate would be the thought processes of mechanical designers as they have sought to satisfy their need for fasteners.

Evolutionist Professor of Geology, Stephen Gould, has stated, "the evolutionary trees that adorn our textbooks have data only at the tips and nodes of their branches; the rest is inference, however reasonable, not the evidence of fossils." (Gould 2)

We conclude that without further evidence of real relationships caused by evolution, the creationists arguments in opposition to a family tree are the more persuasive. The whole tree is not there.

Do the Fossils Support the Concept of Lineage Among Animals?

Is there a clear line of descent from one animal species to another? In order to be valid, a family tree which depicts the lineal descent of each animal from progenitor animals must show a clear fossil record of the lineages connecting the animals. In order to test whether or not the family trees portrayed by evolutionists are valid, I chose what I consider to be one of the best college textbooks on the subject of the biology of animals (Hickman 10), and, for each of the major classification of animals, I searched for the statements which would specify the evolutionary lineage of that group. For each group, I expected there to be a clear statement of its ancestry based on the fossil records. Since there are in our museums 100,000,000 fossils, I assumed that all of the lineages would be amply supported by fossil proofs, possibly with pictures of the fossils. The results of this study, which includes most of the animal kingdom, are recorded in Table 22.1 below. The pages in Hickman are given for the purpose of reference. Please note that the alleged lines of descent are not based on a clear path of the fossils, but rather on uncertain lineages contrived from imagined comparative anatomies.

TABLE 22.1

WORDS IN TEXTBOOK USED TO DESCRIBE ANIMAL LINEAGES

ANIMAL GROUP	WORDS USED TO DESCRIBE ANIMAL LINEAGE	BASED ON FOSSILS	PAGE
Protozoa	Protozoa have left no fossil records; thus our conjectures about protozoan evolution must be based on living forms . . . may be . . . probably . . . origin . . . may be . . . amebas were probably derived . . . origin of ciliates is somewhat obscure.	no	354
Sponges	The origin of sponges dates back to the Cambrian. Another hypothesis . . . Whatever the origin . . .	no	363

What the Fossils Say About Origins

Radially Symmetrical (Hydra-corals, jellyfish)	The origins of the Cnidarians and Ctenophores is obscure. Hypothesis today is . . .	no	383
Acoelomates (Flatworms Roundworms)	Little doubt . . . derived from . . . perhaps . . . Some believe . . . may have given rise . . . characteristics in common . . . It may be probably arose . . .	no	398
Pseudo-Coelomates (worms)	phylogenic relationships . . . are problematic . . . probably distant from . . . share characteristics . . . now consider . . . may well been derived originally from . . . common ancestor . . .	no	411
Molluscs (Clams, octopus)	embryonic cleavage . . . strong evidence . . . It is probable . . . zoologists agree . . . split off from . . . apparently gave rise to . . . may have arisen . . .	no	437
Annelida (Segmented worms)	no hypothesis yet given to explain origins . . . It is believed . . . common ancestor suggesting . . . remote relationship . . . important resemblance . . . more similar to . . . probably evolved from . . .	no	445 454
Arthropods, (Spiders, insects, crabs)	similarities give strong support to the hypothesis . . . line apparently diverged . . . may have come from . . . relationship has long been a puzzle . . . It is believed . . . may have given rise . . .	no	497
Lesser Protostomes, (Sea worms)	probably diverged at different times from . . . background is obscure . . . seem to occupy . . . position . . . between . . . ancestral . . . may have been similar . . .	no	503 512
Echinoderms, (Sea stars, sand-dollars)	share characteristics with . . . early larva is almost identical . . . the early embryogenesis is like . . . shows affinities with . . . some believe . . . maybe similar to . . .	no	530

Protocho- choradates (Hagfishes, lampreys)	very difficult to reconstruct lines of descent . . . reconstructions come from living organisms . . . especially . . . early developmental stages . . .	no	538
Fishes	origin of chordates may never be known . . . descended from an unknown common ancestor . . . may have arisen . . . descended possibly from . . . ancestry shrouded in mystery	no	550 55 571
Amphi- bians, (Frogs, toads, newts)	All evidence points to the lobe-finned fishes as ancestors of the modern amphibians . . . newts and salamanders may have descended from the lepospondyls . . .	no	576
Reptiles (Snakes, dinosaurs, turtles)	Biologists agree that reptiles arose from amphibians 280 mya . . . transition was effected by the . . . amniotic egg . . . this egg may well have developed before . . . turtles have very little change . . . early fossil record of this group (snakes) is poor . . .	no	588 593 594
Birds	the fossil record of birds is disappointingly meager . . . Most paleontologists agree . . . birds probably . . . evolved from a singel ancestor . . . paradoxical that birds . . . should have descended from . . . reptiles. Yet the . . . anatomical affinities . . . evidence . . . kinship	no	605
Mammals (Rodents, bats, cats, whales, apes, men)	In the Mesozoic . . . therapsids (mammal-like reptiles) appeared . . . at some point premammals acquired . . . hair and mammary glands . . . they are believed to have desended . . . numerous resemblances between mammals and reptiles . . .	no	627 629

Obviously in this Hickman textbook, which was authored by zoologists who believe strongly in the theory of evolution, if there were, for any animal group, a clear lineage in the fossil record, this path of lineal descent would certainly have been mentioned. But no such clear cut paths have been identified anywhere in the animal kingdom. Lacking fossil evidence, authors have simply assumed that all animals have descended from ancestors by evolution, and they have attempted to determine lineages by imagining what might have been the anatomies of fossilized animals and then comparing morphological similarities between living animals and these imagined structures, including the anatomies of embryos, juveniles and adults. Again, lacking hard fossil evidence, the Hickman authors have filled their phylogenic descriptions with such words as, "another hypothesis, perhaps, some believe, may have given rise to, probably arose, may well have, it is probable that, probably evolved from, etc."

Evolutionists simply assume that all animals descended from their ancestors by evolution, and they then speculate as to what might have been the paths of such descent based on anatomical morphologies. In fact, the Hickman authors state that "the fossil record suggests that evolution moves in leaps . . . Thus it has seldom been possible to piece together ancestor-dependent sequences from the fossil record." (Hickman 11)

Dr. Colin Patterson, Senior Paleontologist at the British Museum of Natural History, wrote to Luther Sunderland, "You say that I should at least show a photo of the fossil from which each type of organism was derived. I will lay it on the line—there is not one such fossil for which one could make a watertight argument." (Patterson 1)

Creationists note that, even though our museums are filled with 100,000,000 fossils, these fossils do not support the contention that all animals are descended from earlier animals of a different species. Creationists contend that God created each species different and distinct from all others. And the fossil record supports this contention. We must conclude that the fossils support the theory of creation in the matter of the lineages of animals.

Fossils Indicate that New Species Appeared Suddenly, Not Gradually.

If the theory of evolution were in fact the correct explanation for the source of new species, it would be expected that the fossil record would show a long series of minor changes which would gradually transform one species into another. But the fossil record does not support this pattern of events. The fossils, in fact, reveal an opposite characteristic. New species appear suddenly with no evidence of any prior development. The best proof of this is contained in the statements of recognized scientists.

In 1859 Darwin recognized this fact, based on the fossils recovered at that time. Darwin wrote, "The abrupt manner in which whole groups of species suddenly appear . . . has been . . . a fatal objection to the belief in the transmutation of species . . . I allude to the manner in which species belonging to several of the main divisions of the animal kingdom suddenly appear in the lowest known fossiliferous rocks . . ." (Darwin 1)

In 1950, the esteemed German paleontologist, Otto Schindewolf, wrote, "Contrary to the classic theory of evolutionary descent, the . . . designs are not smoothly connected by a long chain of transitional forms . . . but they appear in contrast with one another, set apart by large discontinuities . . . We have found that . . . a family or an order did not arise as a result of continuous modification . . . but rather by means of a sudden, discontinuous direct refashioning." (Schindewolf 4)

In 1953, paleontologist George Simpson stated, "that most new species, genera, and families, and nearly all new categories above the level of families, appear in the record suddenly and are not led up to by known, gradual, completely continuous transitional sequences." (Simpson 1)

In 1986, the Hickman authors wrote, "Most major groups of animals appear abruptly in the fossil record, fully formed, and with no fossils yet discovered that form a transition from their parent groups." (Hickman 12)

Evolutionists have no cogent explanation to account for the sudden appearances of the animals. In fact, this characteristic of the fossil record is diametrically opposed to the conventional evolutionary theory. On the other hand, the sudden appearance of fully-developed animals is exactly what would be expected if God came down to earth, from time to time, to design and construct new animals.

Clearly, the abrupt appearance of new animals in the fossil record supports the creation theory of the origin of animals.

The Fossils Reveal that Animal Species Persist Unchanged for Ages

The fossil record indicates that the general pattern of the existence of animal species is that, after their sudden appearance, they may persist without change, often for millions of years, and then become extinct. David Raup, of Chicago's Field Museum, stated, "Species appear in the sequence very suddenly, show little or no change during their existence in the record, then abruptly go out of the record." (Raup 1)

Schindewolf states that the "horseshoe crab . . . are found today in the coastal regions of . . . North America . . . This crab first appears in the Bunter (Lower Triassic) . . . This species is identical . . . to . . . recent representatives . . . The genus has a life span of . . . two hundred million years." (Schindewolf 5) Based on fossil research performed at the British Museum of Natural History, Craig reveals that species of the following animals have lived on earth, unchanged, for the years indicated: turtles, 275 million years (my); crocodiles, 195 my; silverfish, 395 my; cockroaches, 345 my. (Craig 2) Biologist, Professor Ridley, states, "In the fossil record, stasis—evolutionary lineages that persist for long periods without change—is common." (Ridley 3) Swedish paleontologist Jarvik has stated that, "The main vertebrate stem groups became anatomically specialized some 400 to 500 million years ago and have changed relatively little since then." (Jarvik 1) Also earth scientist, Stanley, has stated, "The record now reveals that species typically survive for a hundred-thousand generations, or even a million or more, without evolving very much . . . After their origins, most species undergo little evolution before becoming extinct," (Stanley 1)

Evolutionists believe that whatever new-feature-producing agent is responsible for causing evolution is acting on the animals at all times, producing new features and new species. If that were true, how could so many species totally escape the effects of this agent and thrive unchanged for millions of years? Creationists, on the other hand, do not believe that there is any new-feature-producing agent that can add elements to the DNA of animals and produce new features. They believe that God designed and constructed each animal species at some time during the long past, and the males and females of each species then mated and produced offspring of the same species, repeating this multiplication until eventually becoming extinct.

The Bible says that "God created . . . every living creature that moveth, which the waters brought forth abundantly, after their kind, and every winged fowl after its kind: and . . . cattle, and creeping thing; and beast of the earth after his kind: and it was so." (Genesis 1:20-24) The term, "after its kind" suggests that each species will persist as that species, and that is what the fossil record confirms, even over periods of time involving millions of years.

Clearly we must conclude that the fossil record reveals the long-term stasis of species, and this supports the theory of creation, and stasis is supported by the Bible.

Mass Extinctions are Revealed by the Fossils

The fossil record reveals that since animals have been on earth, many have become extinct and then new species have appeared. There are two types of extinctions: (1) background extinctions, and (2) mass extinctions. Background extinctions occur to individual species from time to time, at a steady rate, caused by predators, parasites, competition for food, etc. Many animals are becoming extinct today. Mass extinctions involve the sudden disappearance of a large number of animals and a wide variety of species. The fossils reveal that there were two really mammoth mass extinctions and three other very large ones. Ridley says, "At the end of the Permian . . . and the end of the Cretaceous . . . there were mass extinctions . . . the end of the Permian . . . 96% of marine species became extinct . . . , the Cretaceous, . . . 60 to 75% . . ." (Ridley 4) These five mass extinctions occurred, approximately, at the following times: 435 mya, 370 mya, 240 mya, 205 mya, and 65 mya. The largest two were at 240 mya and 65 mya. The trilobites were eliminated at 240 mya and the dinosaurs, at 65 mya. The amphibians were almost eliminated at 240 mya, but several survived, such as frogs, toads and salamanders. (Starr 7)

After each mass extinction there has followed a period during which many new animals appeared, and they then flourished. But often it has been true that small numbers of these animals had actually appeared just before the extinctions. For example, bony fishes appeared, in small numbers, before the mass extinction of 240 mya, but, then, immediately after the extinction, they enjoyed a huge expansion, involving many new species. And, prior to the extinction of 65 mya, small numbers of mammals appeared, but after the 65-mya extinction, which eliminated the dinosaurs, the mammals have really thrived.

Many scientists believe that each extinction was caused by some unusual natural event, such as the asteroid suggested by Alvarez, and they might be correct. However, consider the following scenario which suggests that God may have had something to do with the extinctions. The Bible reveals the characteristics of God. In general, it would appear that God set up the laws of physics and chemistry, and then he created matter and energy. Then he created various specific items, such as our solar system, the earth, and the plants and animals. After each major creation he examined his product and declared that, "it was good." In the first Chapter of Genesis, after creating, or revealing, light, vegetation, stars, sea creatures and land animals, in verses 4, 12, 18, 21, and 25, it is reported that God "saw that it was good." He liked the products of his creation. He apparently enjoyed his creative activities. To him it was fun.

Then, it would appear that God often went away from the earth, but he visited it frequently. He "came down" and he "went up", as reported earlier in this Chapter, but, as time went on, he apparently became disappointed in how things were going, and he threatened or executed violent judgments. After Adam and Eve disobeyed God, he disciplined them, saying to Eve, "I will greatly multiply thy sorrow . . . in sorrow thou shalt bring forth children." And to Adam, he said, "Cursed is the ground for thy sake . . . In the sweat of thy face shalt thou eat bread . . . Therefore . . . God sent him forth from the garden of Eden, to till the ground . . . So he drove out the man." (Genesis 3:16-24) Thereafter men, and possibly also animals, had to compete and work hard for their food.

Prior to speaking with Noah, "God saw that the wickedness of man was great in the earth . . . And it repented the Lord that he had made man on the earth . . . And the Lord said, I will destroy man whom I have created from the face of the earth." (Genesis 6:5-7) Thereafter a huge flood killed much of the life which then existed on the earth.

When men began to build the tower of Babel, God said, "let us go down, and there confound their language . . . and . . . from there did the Lord scatter them abroad upon the face of all the earth." (Genesis 11:7-9) He introduced different languages and then scattered the inhabitants of the earth.

After the children of Israel sinned and made a golden calf to worship, God said to Moses, "I have seen this people, and, behold, it is a stiff-necked people. Now therefore let me alone, that my wrath may wax hot against them, and that I may consume them . . . And Moses besought the Lord his God, and said . . . Turn from thy fierce wrath, and repent of this evil against thy people . . . And the Lord repented of the evil which he thought to do unto his people." (Exodus 32:9-14) God apparently threatened a judgment, but then did not carry out that threat.

The Bible also predicts that there will be a future judgment: "The heavens shall pass away with a great noise, and the elements shall melt with fervent heat; the earth also, and the works that are in it, shall be burned up." (2 Peter 3:10)

These quotations seem to reveal two characteristics of God: (1) After God succeeds in creating something that he likes, it is reported that God "saw that it was good." And (2) apparently when men disobey God, he does not hesitate to execute judgments.

Since it was the same God who created the animals before the time of Adam, as the God who executed great judgments after he created Adam, isn't it logical to conclude that it might have been God who brought about the mass extinctions? Maybe he saw that the dinosaurs were dominating the earth, killing its weaker inhabitants and devouring all of its food supplies, so he experimented with a few mammals, then got rid of the dinosaurs at 65 mya, and then created many more mammals. Possibly this extinction was caused by a huge asteroid, but who threw the big rock? And maybe he saw the dominance of the trilobites prior to 240 mya, and he decided to eliminate them with the great extinction of 240 mya, and he then introduced the bony fish. Obviously, we cannot prove that God caused the mass extinctions, but we can observe that the mass extinctions of animals before the time of Adam, and the severe judgments administered to men after the time of Adam are similar in nature. If we can, correctly, assume that God's characteristics do not change, and if he demonstrated a willingness to exercise judgments after the time of Adam when things didn't please him, isn't it logical to assume that he might have exercised judgments before the time of Adam, such as causing the great mass extinctions?

We can conclude that the fossils reveal that there were several great mass extinctions, and these were usually followed by the creation of many new animals; and, especially in view of the nature of God as revealed in the Bible by his judgments administered to men, it is logical to speculate that, very likely, it was God who was behind some, or all, of these extinctions. This pattern is consistent with the theory of creation.

Do the Fossils Reveal Transitional Species or Gaps and Missing Links?

Since the days of Darwin, even the general public has heard of gaps and missing links in the fossil record; but are they still missing today? According to the theory of evolution, each species slowly evolved from some previous species. If that were true, then, for each species, there should be in the fossil record numerous in-between animals; but the record is notable for its lack of such transitional species. The fossils are similarly missing for genera, families, orders, and classes. Let's examine what the experts say about gaps and missing links. All of the experts we quote in this book are recognized professionals, and most of them are evolutionists.

Paleontologist Stephen Gould said, "All paleontologists know that the fossil record contains precious little in the way of intermediate forms, transitions between major groups are characteristically abrupt." (Gould 3)

Zoologist David Kitts said, " . . . paleontology . . . had presented . . . difficulties . . . the most notorious of which is the presence of 'gaps' in the fossil record. Evolution requires intermediate forms . . . paleontology does not provide them." (Kitts 1)

Geologist Arthur Boucot said, "the inability of the fossil record to produce the 'missing links' has been taken as solid evidence for disbelieving the theory." (Boucot 1)

In his Inaugural Lecture at the University of Queensland in 1980, geologist J. B. Waterhouse said that after 150 post-Darwin years, "the gaps have not been plugged."

The German paleontologist, Otto Schindewolf, said, "Fossil material did not then (in Darwin's day) and . . . does not today meet this challenge (to find the missing links), not by a long shot . . . the individual structural designs stand apart . . . without true transitional forms." (Schindewolf 6)

Paleontologist David Raup said, "Darwin was embarrassed by the fossil record . . . we are now about 120 years after Darwin and the knowledge of the fossil record has been greatly expanded. We now have a quarter of a million fossil species, but the situation hasn't changed much . . . we have even fewer examples of evolutionary transition than we had in Darwin's time." (Raup 2)

The above-quoted scientists are all evolutionists, and their statements don't need additional comment, they speak for themselves. Obviously, with 250,000 fossil species it is possible for an evolutionist to line them all up into what might look like a family tree based on morphological characteristics deduced from the bones, but the transitional animals are all missing.

The creationist model used in this book asserts that God, over the history of the earth, intermittently, and from time to time, came down to the earth to create new animals. The fossil record, regarding transitional animals, gaps and missing links, clearly favors the theory of creation.

What do the Fossils Reveal about Transitional Body Parts?

In my opinion, the total lack of any fossils that represent intermediate stages in the development of body parts of animals, as they presumably evolved from one species to another, is certainly one of the most convincing evidences against the theory of evolution. If invertebrates evolved to become fish with a backbone, where are the fossils showing 25-percent-finished backbones? If fish evolved to become amphibians with pelvis bones and legs, where are the fossils which show that fins changed into partially-developed legs?

All fish and amphibians lay their eggs in water, and the eggs are fertilized after being laid. But reptiles lay shelled eggs on land. Fossils of eggs have been found. Reptile eggs had to be fertilized in the body of the female before the shell was formed. This required an entirely new set of sex glands and instincts. Such changes could only come about by adding thousands of new atoms to the animal's DNA. How could that happen, just by chance? Where are the fossils of the half-formed eggs?

Evolutionists believe that reptiles changed into birds. They believe that reptilian scales turned into feathers. Where are fossils of partially-formed feathers? A bird can only fly if its bones are hollow, thin-walled tubes, reinforced against elastic-instability by internal braces. Are there any fossils of half-finished bird bones? Evolutionists believe that reptiles also evolved into mammals. Mammals have three bones in their ears; reptiles have only one. Are there any fossils of intermediate animals with 1.5, 2, or 2.5 bones in their ears? Reptiles have four bones in the lower jaw; mammals have only one. Are there any intermediate fossils with 3.5, 3, 2.5, or 2 jaw bones? Are there any bones in your body which are partially-developed portions of some new feature? Are you evolving? All of these fossils of partially-developed new features are missing! These half-finished new parts are not just in short supply, they are absolutely and totally missing!

Creationists believe that throughout the history of animals on earth, God, from time to time came down to earth and he designed and constructed new animals, each one complete and fully-developed; God created no half-finished parts. This characteristic of the fossil record, the total absence of any partially-finished animal parts, clearly favors the theory of creation.

Fossils Show that Bones of Different Animals may have Similar Designs.

The fossils show that the bones of a wide variety of animals often have remarkably similar designs. Figure 22.5 shows the bones of the limbs of several different animals. Note that they all have somewhat similar humerus, ulna, radius and wrist bones. Evolutionists assert that such similarities support the contention that they all descended by evolution from some common ancestor. However, if that were true, then the fossil record should contain numerous examples of the intermediate transitional parts. But not one such fossil has ever been found.

Creationists assert that these similarities simply reflect the fact that when a designer finds a good design, he applies it in many applications. Since there is no fossil evidence that any one of these limb bones evolved from one species to another, or from any common ancestor, and since there is no known new-feature-producing agent that could change thousands of atoms in the DNA molecules of these animals, and since the explanation for these similarities offered by the creationist is reasonable, we must conclude, again, that these design similarities favor the theory of creation.

Figure 22.5 Bones of Different Animals have Similar Designs

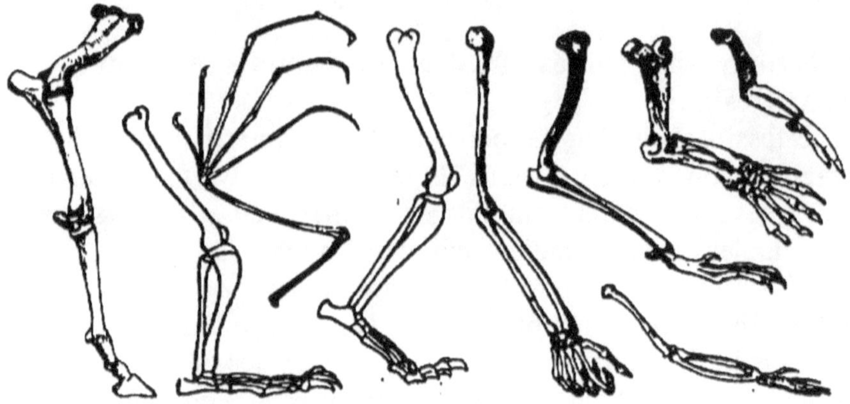

Horse Bear Bat Tiger Man Monkey Dog Manatee Penguin

Fossil History Shows Increase in Complexity of Animals.

Evolutionists claim that the history of the fossils of animals shows a pattern of ever-increasing complexity, from the protozoa to modern humans, and that this supports the theory of evolution. The following assertions are probably accurate descriptions of the history of animal complexity over time.

(1) All animals are very complex.
(2) It is difficult to rank animal complexity.
(3) There probably has been a trend of increasing animal complexity.
(4) The trend has not been uniformly upward for all animals.
(5) Increased complexity does not necessarily increase survival capability.

Each of these assertions will be discussed in more detail.

All animals are very complex. Even single-celled animals, such as protozoa, are controlled by a DNA molecule that may have as many as 68,000,000 carefully-arranged atoms of carbon, oxygen, hydrogen, nitrogen, and phosphorus; and these animals have all of the organelles necessary to carry on life. Anderson and Druger state that, "Protozoa perform all of these activities within the small space of the single cell that forms their body . . . These processes include movement, ingestion and digestion of nutrients, excretion of waste matter, respiration, and maintenance of internal chemical concentrations, growth and reproduction." (Anderson 1) Obviously, even these single-celled animals are extremely complex. Multi-celled animals are even more complex, and the strange creatures of the Cambrian Explosion were unbelievably complex. Modern humans are so complex that, even after the expenditure of billions of dollars on research, the medical profession is still trying to figure out how the human body works. It is reasonable to conclude that all animals are extremely complex.

It is difficult to rank animal complexity. Probably because all animals are very complex, and because there is such a wide variety of animals, it is difficult to rank the complexity of animals. Which is more complicated, a bee or a bird? I marvel at the clever design of the insect's wings described in Chapter 12. Is a mouse more complicated than a spider? Is a house fly more complicated than a trilobite? Complexity is hard to evaluate accurately, but certain generalizations can be made.

Has animal complexity increased with time? The animals that have existed on the earth are identified in the Section on Lineage in this Chapter. They include: protozoa, trilobites, sponges, jellyfish, corals, anemones, worms, snails, clams, squid, spiders, insects, fishes, amphibians, reptiles, birds, and mammals. Certainly mammals are more complex than protozoa, and there does seem to be a general trend toward more complex animals with time. Valentine has plotted complexity vs number of cell types, and he concludes that complexity of animals has increased steadily and nearly linearly from the origin of the metazoa to the present. (Valentine 1) So, we will conclude that this trend of increased complexity has, in fact, occurred.

Complexity has not increased uniformly. Although there probably has been a general increase in the complexity of animals over the past 540,000,000 years, this does not imply that each new animal is very similar to some previous animal, but with some small amount of complexity added. On the contrary, in previous sections of this Chapter we have shown that new and different animals have appeared suddenly in the fossil record, and there have been gaps and missing links, and there are no transitional species or transitional animal parts. Thus, although the general increase in complexity is a discernible trend, it is certainly not a clearly demonstrable characteristic of the fossil record that applies to all animals at all times. Geologist Steven Schafersman has stated that, "Although we observe a general tendency for life to evolve from simple to complex, . . . the record shows simple and complex organisms living side-by-side in every geologic age." (Godfrey 3)

Does increased complexity necessarily increase survival capability? Which is more complex, an insect or a dinosaur? Insects appeared about 300 mya, while dinosaurs appeared later, about 200 mya. But we still have insects with us, and the dinosaurs have become extinct. Many of our most advanced and complex animals are becoming extinct today. Clearly, complexity does not necessarily produce survival value, but survival value is the quality most needed by the theory of evolution.

Can the theory of creation explain increased complexity? Consider the following scenario. The God of the Bible, from time to time during the past 540,000,000 years, came down to earth and created animals. He, at first, created simple unicellular animals such as the protozoa. He then created multicellular animals, and then, as the years progressed he continued to come down when it pleased him, and he created a few new animals on each visit, until all of the 40,000,000 species of the animal kingdom had been designed and built. If, on average, he created 40,000,000 animals in 540,000,000 years, he would have had 13.5 years in which to design each animal. I can imagine that this design and construction of animals was fun for God. All engineering designers derive pleasure from designing clever machines. And the Bible repeatedly states that after God had designed something, he said that "he saw that it was good." I can also imagine that God, after enjoying the creation of one animal, said to himself, "It would be fun to create a more complex animal." This process of creating animals, and enjoying the gradual increase in their complexity, explains the fossil record we find.

Can the theory of evolution identify any natural force that might have the capability of changing one animal into a more complex one? It might be generally true that, by natural selection, the more complex animals might be more likely to survive. But natural selection is not a new-feature-producing agent. It just differentially selects what already exists.

Without any answer to the above question, and based on a logically conceived concept of how God might have acted, we can conclude that the matter of the increasing complexity of animals favors the theory of creation.

Do the Fossils Indicate that Humans Evolved from Ape-Men?

Evolutionists believe that modern man and modern apes both evolved from some common ape-man ancestor. Our task is to study the fossils and determine the accuracy of this contention. Evolutionists use the word, "hominid" to refer to humans and their evolutionary ancestors. So, how many hominid fossils do we have to study? There is a myth afloat which assumes that there are only a few hominid fossils, and they would all fit in one coffin. Actually, the *Catalogue of Fossil Hominids* asserts that, as of 1976, a total of 3998 fossil individuals had been discovered. Fossil researcher, Marvin Lubenow, believes that, as of 1992, there are about 6000 such fossils. (Lubenow 2) These fossils are locked inside of reinforced concrete vaults in museums. Many paleontologists have never seen an original fossil. Information about them is obtained from cast replicas and from published literature.

Scientists frequently argue about the dating of the fossils, and about the identification and classification of the fossils. In some cases the agreed-upon dating and identification has changed with time. For example, big names among fossil types would include: Nebraska Man, Java Man, Piltdown Man, Peking Man, Heidelberg man, Neanderthal Man, Lucy, Ramapithecus, Australopithecus, etc. Some of these have proven to be as originally claimed, and others have not. Nebraska Man, consisting of one tooth, proved to be an extinct pig. Java Man was probably an extinct gibbon. Piltdown Man was an actual hoax. But these unfortunate examples are not typical of the vast majority of the fossils, although each new fossil usually stimulates some debate among the experts. In this book we are going to accept the dating and identification of the fossils as agreed upon by the majority of the scientists. We will concentrate on trying to determine who designed and constructed these hominid animals, including humans. How did they come into existence, by evolution or by deliberate creation?

Pictures of ape-men are nothing but the products of the imaginations of artists.

Scientists have organized the 6,000 hominid fossils into groups, such as Australopithecus africanus, Homo erectus, archaic Homo sapiens, etc. These categories are recorded on Figure 22.6, which shows the time period during which each group is thought to have existed on earth. This chart is based on time periods suggested by evolutionists, and it should be noted that the ending of Group A occurs at the same time as the beginning of Group B, and C begins when B ends, and so forth through ABCDEF and G. This means that, according to the evolutionists, A could have evolved into B, B into C, C into D, etc., until G were reached. The dotted lines on the Figure depict this progression. Thus, according to this chart, Australopithecus afarensis could have evolved, eventually, into modern humans.

The animal groups represented by X, Y and Z are other hominids which many evolutionists believe were not in the direct line of ancestors of Homo sapiens.

But Professor Lubenow has extensively researched the hominid fossils and he contends that the datings of the fossils that make up the convenient path of ABCDEFG in Figure 22.6 are substantially in error. He has prepared equivalent data for the representative fossils entered in Table 22.2. (Lubenow 3) This Table lists specific fossil specimens, arranged vertically by their dates of existence, and placed, horizontally, into four columns, each of which represents a hominid group. The first column contains Homo sapiens (G of Figure 22.6); the second, Neanderthals (F) and archaic Homo sapiens (E); the third, Homo erectus (D); and the fourth, all of the Australopithecines. If the data of Table 22.2 were plotted as in Figure 22.6, the result would be four multimillion-year lines. Please note that the oldest date for entries under archaic Homo sapiens in Table 22.2, is nowhere near the most recent date of the Homo-erectus entries, and the oldest date of the Homo sapiens is a staggering 4 million years older than the most recent archaic-Homo-sapiens. The same

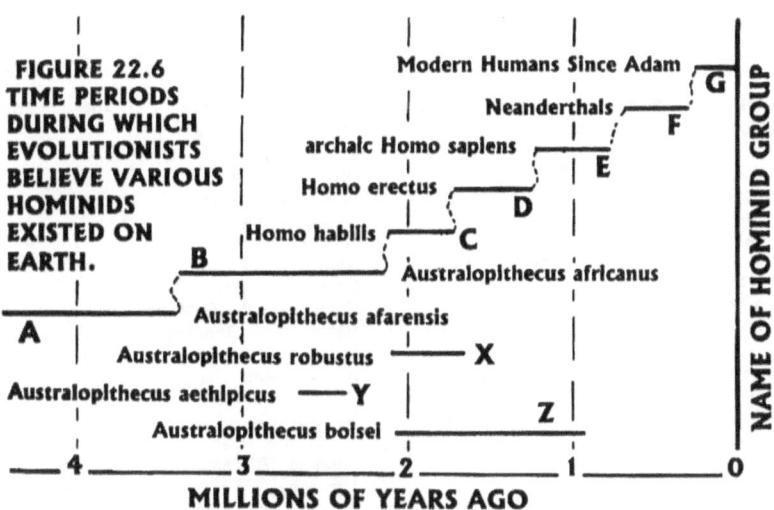

FIGURE 22.6
TIME PERIODS
DURING WHICH
EVOLUTIONISTS
BELIEVE VARIOUS
HOMINIDS
EXISTED ON
EARTH.

Modern Humans Since Adam

Neanderthals

archaic Homo sapiens

Homo erectus

Homo habilis

Australopithecus africanus

Australopithecus afarensis

Australopithecus robustus

Australopithecus aethipicus

Australopithecus boisei

NAME OF HOMINID GROUP

MILLIONS OF YEARS AGO

observation applies to the Homo erectus group and the Australopithecines. This means that large segments of all four groups were contemporaries, and from this it can be concluded that the animals of one group could not have evolved into the animals of another group, and the chain of evolutionary progression suggested by Figure 22.6 could not have occurred.

Table 22.2. Fossil Chart. Prepared by Marvin Lubenow.

modern *H. sapiens*	archaic *H. sapiens*	Homo erectus	Australopithecines
Springbok Flats	Cape Flats/Dire Dawa	Mossgiel/Cossack	
Predmosti/Brno	Tangier/Eliye Sp.	Lake Nitchie	
Lake Mungo/Keilor	Dar es Soltan	Java Solo people	
Cro-Magnon/Niah	Haua Fteah/999	Kow Swamp/Cohuna	
Bacho Kiro/Liujiang	Florisbad/L. H. 18	Coobool/Talgai	
Jebel-Qafzeh	La Chaise/Azych	WHL 50/Témara	
Wadjak/Singa/Krapina	Krapina D/Eyasi	Narmada/Xujiayao	
Border Cave/Skhūl	Montmaurin	Cave of Hearths	
Klasies River/Mumba	Klasies River	Hazorea/L. H. 29/Mapa	
Fontechevade/Omo 1	Omo 2/Jebel-Irhoud	Hexian/Wadi Dagadlé	
	Bilzingsleben/Garba	DingCun/Changyang	
Pontnewydd	Castel di Guido	Rabat/Peking Man	
Rhodesian leg bones	Rhodesian Man skull	Sidi Abderrahman	
	Saldanha/Atapuerca	Kapthurin/Salé/Yunxian	
	Swanscombe/Steinheim	Jinniu Shan/Dali	
.5 m.y.a.	Bodo/Arago 21	Thomas 1&3/Arago	
Java Man femur	Vértesszöllös	Luc Yen/Lang Trang	
Sondé molar	Ndutu	Java Man skullcap	
		Peking Man/ O. H. 23	
		Lainyamok/Lantian	
		Ternifine/Yunshien	
		Java (Sangiran)	
	Petralona	Olduvai Hominid 12	
	Mauer	Olduvai Hominid 28	Taung
		Olduvai Hominid 22	
		Yayo	
		Yuanmou	
		Dmanisi	
		Olduvai Hominid 2	
1 m.y.a.		Lantian	
		Olduvai Hominid 29	
		Olduvai Hominid 51	
		Gomboré II	
		Olduvai Hominid 9	
		Olduvai Hominid 36	
		Sambungmachan	
		Omo L-996-17/O.H.15	
		KNM-ER 992, 803	
		KNM-ER 3883, 737, 820	
		KNM-WT 15000	
Koobi Fora prints		KNM-ER 1808, 730	
Gomboré IB-7594		KNM-ER 3733, 1507	
KNM-ER 813/O.H. 48		SK-15, 18a, 18b	
Olduvai Structure		SK-84/SK-847	
2 m.y.a.		SKX-5020	
KNM-ER 1472		Java (Djetis)/Damiao	
KNM-ER 1590		KNM-ER 3228	
KNM-ER 1470, 1481			
3 m.y.a.			Lucy
Laetoli footprints			
4 m.y.a.			
Kanapoi humerus			
5 m.y.a.			

Vertical labels: Neandertals; Homo habilis (flawed taxon); A. africanus; A. robustus/boisei; A. Afarensis

240

In spite of the above, some evolutionists still believe that modern man evolved, first from some animal such as an ape, then through some transitional species such as an ape-man, and then to modern man. First, let's dispel the myth that has been imposed upon the general public by evolutionists, through pictures of ape-men. These bent over, heavy-browed, hairy creatures always look more like apes than men, and they are all strictly figments of the imaginations of artists. Secondly, let us ask what is the scientific name for the transitional ape-man? He has never been named. The reason for this is because no ape-man fossils have ever been found. Then, according to evolutionary theory, if an ape evolved into an ape-man, the survival capability of the more advanced ape-man should be greater than that of the ape. But the ape is still with us and the ape-man has disappeared. It is probably true that no ape-man ever existed! If evolution did not produce the hominids and their fossils, then who did? And what might have been the details concerning their origins? Consider the following.

As you might surmise, I believe that God created the hominids, but who were all of the individuals who appear to have lived on earth between 4 mya and the present, and what might have been their relationship to modern man? In an effort to answer these questions, let's assume that God was a design engineer, similar in many respects to Henry Ford. Being a design engineer myself, I can explain to you that whenever a complex new product is to be designed and developed, it is a common practice to produce one or several prototypes. These can then be subjected to testing and evaluation which might bring forth information that could be applied when later editions are designed and built. Table 22.3 shows the prototypes that Henry Ford designed before he finally settled on the Model T Ford as his production Model. He then produced 15,007,033 Model T's in the United States.

TABLE 22.3

PROTOTYPES AND PRODUCTION VEHICLES PRODUCED BY
HENRY FORD DURING THE PERIOD 1896 THROUGH 1927

NUMBER OF CARS	YEAR	MODEL	PROTOTYPE OR PRODUCTION
1	1896	Quad 1	Early Prototype
1	1896	Quad 2	"
1	1901	Racer 1	"
2	1902	Racer 2	"
1708	1903	A	Later Prototype
1695	1904	B,C,F	"
8729	1906	N,R,S,K	"
10202	1908	T	Production
32053	1910	T	"
78440	1912	T	"
308213	1914	T	"
Continued Through	1927		

Please note the general pattern in Table 22.3. During the first ten years, prototypes were designed, built and tested. Then, in 1908, the development period was completed and production Model-T Fords began to be produced. Their production soon expanded, and, in 1914, 308,213 Model-T cars were built.

The fossil record of the animals shows the same pattern. Figure 22.7 shows the pattern of Number-of-Animal-Families vs Passage-of-Time for mammals, birds, reptiles, fishes, and protozoa. In every case there has been a prototype-development period, indicated by a slender vertical stem, followed by a greatly widened production period shown as we move downward on the plots with the passage of time. Let's consider mammals in greater depth, as an example. Look at the fossil record of the mammals in Figure 22.7. Apparently God produced prototype mammals for some time before 65 mya. Then, the mammals really began to flourish, and they have continued to prosper until the present time. The same pattern holds for the other animal groups. Similarly, isn't it conceivable that God, the great engineer, experimented for a few million years with a wide variety of erect, bipedal, upright-walking, large-

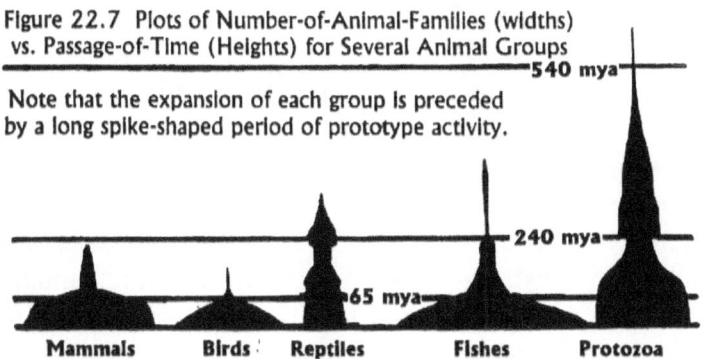

Figure 22.7 Plots of Number-of-Animal-Families (widths) vs. Passage-of-Time (Heights) for Several Animal Groups

Note that the expansion of each group is preceded by a long spike-shaped period of prototype activity.

540 mya

240 mya

65 mya

Mammals   Birds   Reptiles   Fishes   Protozoa

brained prototype creatures, who were very similar to, but not identical with, the ultimate production model, modern man, Homo sapiens? When I suggest that God might have made prototype animals, I certainly don't want to give the impression that I don't think God was smart enough to design animals without making prototypes. I just think he did it because it was fun. He enjoyed it, just as any design engineer enjoys his work. The fossil record seems to indicate that God did, indeed, make prototypes. Why he did it, is another matter.

According to Lubenow's research, recorded in Table 22.2, there have been many of these prototype hominids that lived during the past 4 million years, whose bones cannot be distinguished from the bones of modern man; and many others, such as archaic Homo sapiens, Homo erectus and Australopithecines, whose bones are only slightly different from those of Homo sapiens. Some of these hominids had brain sizes even greater than the brain size of modern man. But brain size is not the only criterion for measuring intelligence. Brain quality is what is important. I think God reserved modern man for the ultimate endowment, the brain of the Homo sapiens.

I think it is logical, consistent with the usual pattern of the design and development of any complex machine, and consistent with the fossil records of all major animal groups, to conclude that, after 4 million years of experimenting with and enjoying prototype hominid creatures, God finally designed and constructed his ultimate production machine, modern man, probably about 8000 years ago. He made Adam and Eve. And in man he put a new, high-quality brain, and endowed man with all of the other marvelous features that distinguish Homo sapiens from the other animals.

Is there any hard evidence to support the above scenario? You probably don't think there is; but, believe it or not, I have discovered some remarkable facts. Consider the following. One of the most recent and most interesting prototype hominids is the Neanderthal man. The Neanderthals arrived suddenly about 200,000 ya, and most of them became extinct about 30,000 ya. Fossil remains of about 300 of them have been discovered. These hominids had a brain which was about 10% larger than the average brain of a human. They had a low, broad, elongated skull, with large heavy brow ridges. Their skull was thick and their trunk and limb bones were massive. Their fossils suggest a people of incredible strength and power. Professor Trinkaus wrote, "Neanderthals . . . bones suggest a strength seldom attained by modern humans". (Trinkaus 1) Geist stated that "Neanderthal was far more powerful than modern humans . . . the skeleton reflects a supremely powerful musculature." (Geist 1) Scientists do not know where Neanderthals came from, or where they went; and they do not fit in the family tree. Stringer says their beginnings 200,000 ya are as much a mystery as is their rapid disappearance at about 34,000 ya. (Stringer 1)

However, Lubenow suggests that, "there is evidence that the Neanderthals persisted long after their alleged demise . . . skull Amud I from . . . Israel . . . was found . . . just below . . . layer BI . . . The radio-carbon date for Upper BI is 5710 ya . . . it could mean that Neanderthal . . . persisted until quite recently." (Lubenow 4) Do you comprehend what this dating means? It means that the Neanderthal hominids were probably in existence on earth when Adam was created!

The first Neanderthal fossil was discovered in Germany in 1856. Recently, just prior to 1997, Dr Svante Paabo, of the University of Munich, has been performing research on the bones of this hominid. Apparently the bones of this skeleton are so recent that they did not completely fossilize, and some original bone material has remained, and it was possible for Dr. Paabo to get DNA samples from these bones. Dr. Paabo tested this DNA and compared it with human DNA, and he determined conclusively that the Neanderthal man was clearly not a human. Neanderthal was a distinct species. I have found a newspaper account of this earth-shaking discovery which I am reprinting here as Figure 22.8.

Figure 22.8   DNA Proves that Neanderthals Were Not Humans

Columbus Dispatch
July 11, 1997

LONDON — There's no sign of Neanderthals lurking in our family tree, researchers declare, saying that DNA from a Neanderthal skeleton is helping resolve a great debate in human evolution.

Distinct genetic differences indicate the Neanderthals were a different species than the early humans who swept them aside in Europe and western Asia.

Results of a five-year study, led by Dr. Svante Paabo of the University of Munich in Germany, were published in the British journal *Cell.*

Paabo came to the conclusion after comparing modern human DNA with a DNA sample taken from the first Neanderthal skeleton ever found, in 1856.

Neanderthals — large, thick-boned, with heavy brows — lived in Europe and western Asia from 300,000 years ago, dying out about 270,000 years later.

For the latter part of that period they clearly co-existed with modern humans, but the relationship between the two groups has been debated since the first Neanderthal was discovered.

Early humans and Neanderthals may have interbred, as some scientists contend, with modern Europeans being descended from both; or the two hominid lines may have remained distinct, with humans displacing and probably slaughtering their rivals.

Paabo concluded that the DNA tests prove Neanderthals were a distinct species, an evolutionary dead end that contributed nothing to the modern gene pool.

On the subject of who created man, what this means is that evolutionists cannot claim that humans evolved from Neanderthal hominids. Modern humans and Neanderthals probably co-existed for a time. The sudden appearance of the Neanderthals just before the creation of humans 8000 years ago is consistent with the prototype theory that I am proposing. I suspect that all of the other hominids represented by the fossils identified in Table 22.2 were also God's prototypes that he created prior to designing modern man.

Finally, believe it or not, I have discovered that the Bible may have something to say about the Neanderthal men. In Genesis, the Bible says, referring to the time period after Adam and before Noah:

> "And it came to pass, when men began to multiply on
> the face of the earth, and daughters were born unto them,
> That the sons of God saw the daughters of men that they
> were fair; and they took them wives of all whom they

> chose . . . There were giants in the earth in those days;
> and also after that, when the sons of God came in unto
> the daughters of men, and they bore children to them,
> the same became mighty men who were of old, men of
> renown." (Genesis 6:1-4)

A "son of God" would surely have to be a creature made by God, but yet these "sons of God" were described as different from men. But yet they apparently could have sex with human women, "the daughters of men." Also, they are described as very large and very strong. They were "giants" and their offspring are called "mighty men." I can come to no other conclusion than that the Bible is actually referring to the Neanderthal people.

It is further interesting that right after the above account God said,

> "And God saw that the wickedness of man was great
> in the earth, and that every imagination of the thoughts
> of his heart was only evil continually . . . and I will cause
> it to rain upon the earth forty days and forty nights; and
> every living thing that I have made will I destroy from off
> of the face of the earth." (Gen 6:5, 7:4)

Is it possible that one reason God brought about the flood was to get rid of the prototype Neanderthals and other non-human hominids that he had made?

I believe it is very probable that the Neanderthal men and all of the other hominids that existed on the earth during the 4 million years prior to the creation of Adam were God's prototype creations, and they preceded the creation of modern man, just as Henry Ford's Models A,B,C,F,N,R,S and K preceded his final production car, the Model T. I think that the concept of prototypes is very reasonable. It is consistent with the design process, itself. It is consistent with the patterns of fossil records for most animals. It is consistent with the facts concerning the Neanderthals. It explains the purpose of the creatures which produced the 4 million years of human-like fossils on earth. It distinguishes between modern men and all of the other hominids. It is consistent with the Biblical account of the Neanderthals. It is consistent with the Biblical account of Adam and Eve. It is consistent with the dating of Adam and Eve at 8000 ya.

It is consistent with Bishop James Ussher's chronology, which can give dates between 6000 ya to 10,000 ya. We've used 8000.

So, who were the ape-men? They didn't exist! But 4 million years of bipedal upright-walking hominids did exist. They were the prototypes God made just prior to his creation of man. The general conclusion to which we must come relative to what the fossils say about the origin of man, is that the fossil record supports the theory of creation.

Conclusions Concerning What the Fossils Say About Animal Origins

Let's summarize what we have learned from the analyses of this Chapter. Although the fossils can be arranged into something that looks like a family tree, the tree is incomplete, the branches are not connected. The tree concept makes an attractive display, but it cannot withstand rigorous scrutiny. The animals appeared in the fossil record suddenly, not gradually. They persisted unchanged for long periods of time, until becoming extinct. The mass extinctions could well have been caused by God; being consistent with his demonstrated willingness to bring about judgments. There are no clearly documented transitional species or transitional body parts. There are gaps and missing links. Similar designs in different animals simply indicate that if a designer finds a good design he is likely to apply it in many products. A good designer would be expected to continually improve the quality and complexity of his designs. And our ancestors called the ape-men did not really exist, those creatures were the prototypes that God made before he designed his ultimate production model, Adam and Eve and modern man. The concept of hominid prototypes is remarkably supported by the fact that we have actually found that the non-human Neanderthal man is mentioned in the Bible.

Every one of the analyses of this Chapter has shown that the fossils support the theory of creation more than the theory of evolution. This should not be surprising if the theory of creation is, in fact, the correct theory. I believe that all of these analyses are based on facts or the objective opinions of experts, most of whom are evolutionists. Our general conclusion, then, must be that the fossil record clearly supports the theory of creation.

The Cambrian Explosion

CHAPTER 23. THE AFFIDAVITS
OF EVOLUTIONISTS

Sources of Affidavit Material

An affidavit is a sworn written statement presented by a potential witness in a law suit, which summarizes the testimony of that witness. The format of this book is that of a law case in which I, the Author, am serving as an expert witness, and my assignment is to determine who created the animals, who caused the Cambrian Explosion. Other witnesses in this law case should include evolutionists, who also have opinions on who created the animals. And, one of the tasks of a good expert witness is to learn what are the facts and opinions to which the other witnesses might testify. In an effort to learn what evolutionists really believe on this subject, and why they believe as they do, I sent a questionnaire to more than 1000 college professors and researchers, high-school science teachers, and curators of natural-history museums, all over the United States. The composite affidavit of this Chapter is based on the responses from this questionnaire, plus the statements and writings of evolutionists which have come to my attention during the past several decades.

To What Will the Evolutionists Testify?

About three-fourths of the questionnaire respondents stated that they believed in the theory of evolution. About half of them did not believe in the existence of God. The basic beliefs of evolutionists were confirmed by the questionnaire, namely, that the natural forces of the earth, alone, created the animals. A more detailed statement of the theory of evolution is given in Chapter 22.

Most evolutionists defend their positions on this subject with great enthusiasm and vehemence. Oxford scientist, Richard Dawkins, said, "It is absolutely safe to say that if you meet somebody who claims not to believe in evolution, that person is ignorant, stupid, or insane." But, in the previous Chapters of this book, I think we have shown that the theory of evolution, to be valid, needs some effective New-Feature-Producing Agent (NFPA), and none has been found; and the fossils do not support the theory of evolution. Why is it, then, that evolutionists are so resolutely convinced that evolution is a fact?

Why do Learned People Believe in Evolution?

I think there are probably six major reasons why learned people, and the general public, believe in evolution. They are the following:

(1) The idea of evolution sounds plausible.
(2) We are all indoctrinated on the subject of evolution by educators and the media from the cradle to the grave.
(3) The indoctrination has been so effective and so complete that few people have been motivated to investigate any alternatives on their own.
(4) Atheists are compelled to embrace evolution.
(5) Some people view the beliefs of some creationists as extremist and unscientific, and hence they discredit the views of all creationists.
(6) The matter is often viewed simply as a conflict between science and religion.

Each of these will be discussed in further detail.

The Idea of Evolution Sounds Plausible.

I could arrange a group of fasteners, such as tacks, nails, screws and bolts, into an array that resembled a family tree, and then tell you that they all evolved one from another. I could do the same with automobiles from the past 100 years. But you know that these did not evolve; they were designed by humans. However, animals are different from the above for three reasons: (1) Animals actually give birth to their successor offspring; (2) the offspring can be different from their parents in a wide variety of ways; and (3) new animals have appeared on earth during the past 540,000,000 years, and they have become increasingly more complex. If you put these observed facts together, and then offer the explanation that they all evolved one from another, the idea sounds plausible. However, it also sounds plausible that the earth is flat, and the sun rotates around the earth. So, these matters required deeper investigation in order to arrive at the truth.

We Are All Indoctrinated on Evolution From the Cradle to the Grave

When the scientists of the world learned of the theory of evolution from Darwin and others, they eagerly embraced the idea because it sounded plausible, and because it enabled them to eliminate God and the supernatural from their science. They soon became convinced that evolution was a fact, and, through their publications and in their roles as esteemed scientists and educators, they indoctrinated their followers and the general public with the proposition that evolution was a fact. Before we learned about DNA, and before much was known about genetics, or about the fossils of animals, there was little scientific knowledge available to refute the theory of evolution; and it flourished for decades, until almost all scientists and educators became so thoroughly indoctrinated with the theory that they assumed it to be a fact. Today, students in grade schools, high schools and colleges are taught that evolution is a fact. Also, inbreeding in the universities exists. Graduate students must be evolutionists. As a result, all of us have been thoroughly indoctrinated in the concept of evolution.

Few are Motivated to Investigate Alternatives.

The indoctrination ingrained in the minds of the scientific community and the general public on the subject of evolution has been so effective and so complete that few people have been motivated to investigate any alternatives. They just assume that the specialists in the field obviously know more than they do on the subject, and so they accept their opinions. They do not learn that most scientists have never observed a beneficial mutation, nor have scientists observed millions of atoms being added to the DNA of an animal. They never learn that no scientist has ever found fossils of half-finished new parts of animals. They do not note the significance of the fact that much of the research of evolutionists consists of, first, assuming that evolution is a fact, and then, by comparing assumed morphologies of extinct animals with modern animals, they assign evolutionary descents. They never learn of the theory of creation, as explained in this book. The average young scientist or layman just accepts what the esteemed evolutionists say, and they never study the matter further, on their own. Just as children are often indoctrinated by their parents in such matters as politics, economics and religion, so have young scientists and the general public been indoctrinated by the evolutionists. They have been taught that evolution is a fact, and they have just passively accepted that as the truth.

Atheists are Compelled to Embrace Evolution.

There are only two widely-held views on the subject of who created the animals, evolution and creation. Some try to wed the two, such as by asserting that God used evolution to create the animals, but so little evidence exists to support these other theories that they will not be considered here. The theory of creation asserts that God created the animals. If a person does not believe that God exists, he cannot believe in the theory of creation. If there are only two theories, then the atheist is compelled, by default, to believe in evolution.

You will recall that about half of those who responded to my questionnaire declared themselves to be atheists. Among scientists, I suspect that the percentage is much higher. I have always been interested in why some people do not believe that God exists. The questionnaire, plus other observations, provide some answers to this question. Many atheists say they don't believe that God exists simply because they have never seen or heard God, and no aspect of God has ever been detected by any of their senses. Also, they cannot conceive of any being who could violate natural laws and have supernatural capabilities. They also believe that God is related to religions, and, since they think that religions are filled with myths and factual errors, they don't want any association with religions. They think that religious people, generally, are of low intelligence and are poorly educated.

Thus, many people believe in evolution simply because they do not believe that God exists, and the only alternative is the theory of evolution. These evolutionists have not chosen to believe in the theory because of the weight of evidence in support of it; actually, they would be compelled to believe in it even if there were no evidence in support of it at all. Many people are evolutionists simply because they are atheists.

The Views of Some Creationists are Characterized as Unscientific.
Some people may tend to embrace the theory of evolution because they cannot accept the views of some creationists who take positions and make statements which they believe are unscientific and false. For example, some creationists proclaim that the universe is less than 10,000 years old, all of the animals were created by God in one 24-hour period some 6,000 to 10,000 years ago, and that most of the fossils of the earth were produced some 4,000 years ago by the flood mentioned in the Bible. Since most scientists disagree with these statements, they discredit all of the other views of these creationists, and they give no credence to the views of any creationist. Their rejections of some of the views of some creationists ignores the major contention of all creationists, namely, that the complexity of all animals leads inexorably to the conclusion that the animals had to have been designed and constructed by some supernatural being, and the only being capable of qualifying for this task is the God of the Bible. But since some scientists discredit the beliefs of some creationists, they tend to reject all creationist ideas, and choose, instead, to embrace the theory of evolution.

The Two Theories are Viewed as a Conflict Between Science and Religion
Scientists and the general public have often viewed the two theories as a conflict between science and religion. You, no doubt, are familiar with the widely-accepted statement that the theory of evolution is wholly scientific and the theory of creation is essentially a religious doctrine. Let me shock you by asserting that these roles actually should be reversed!

We certainly do not want to become embroiled in an argument which is based solely on semantics, but to pursue this contention at all, requires that we define several terms. Science may be defined as knowledge based on observation, identification, description, or experimental or theoretical investigation of natural phenomena, or any class of phenomena. The objective of a scientist is to discover knowledge. A religion is a particular integrated system which expresses belief in, and reverence for, a superhuman power recognized as the creator and governor of the universe. This superhuman power could be nature itself, or the God of the Bible. Religions are usually thought to be based, in part, on faith. Faith is a belief that does not rest on logical proof or material evidence. We should recognize that both evolutionism and creationism are theories. A

253

theory is a system of assumptions devised to explain the nature of a set of phenomena. With these definitions in mind, let's examine the theory of evolution and the theory of creation to determine which one is largely science and which one is largely religion. Our judgment should be based on which theory best fits the facts.

For evolution to be a science rather than a religion it would need to be based on some hard facts, such as knowledge of some New-Feature-Producing Agent (NFPA) that scientists could observe in action. Darwin spoke mostly about natural selection, which we all agree, can be observed to act in nature. But he tended to ignore the need for an NFPA to provide new features that could be acted upon by natural selection. After Darwin, emphasis was placed on Mendelian genetics, until we learned that genetic variation could produce great variety within a species but it could not produce a new species. More recently, emphasis has been placed on mutations, until our knowledge of DNA has shown that randomly caused beneficial mutations, statistically, could not account for the creation of significant new features. Our Chapters in this book demonstrate clearly that there is no basic mechanism which can explain evolutionary development. Furthermore, as explained in Chapter 22, the fossil record does not support the theory of evolution. We must conclude, then, that the factual, scientific basis for the theory of evolution really is missing.

Therefore the theory of evolution must be a religion, based on faith. How much faith does it take to believe that a protozoan could, by chance, arise out of sea-water? How much faith does it take to believe that reproduction by hard-shelled eggs developed by chance? How much faith does it take to believe that lizards evolved into birds, or that a worm could turn into a butterfly? In my opinion, it takes a lot of faith, and religious zeal, to believe in such miracles, and there are no scientific facts to support the theory. Hence, the theory of evolution would appear to be better classified as a religion than a science.

With respect to the theory of creation, it depends greatly on a belief in God. Do you find it hard to believe that God exists? Do you believe that George Washington existed? You have never seen George Washington, or heard him speak. But George Washington has been written about in books and documents. Men have seen him and talked to him, and they have written about these experiences. George Washington led his people in battle against the British, and prevailed. But God has been written about also. There are 66 books that were written about God, when he was a frequent visitor to the earth. How many books were written about George Washington in his hey-day? And God led his people against the Egyptians, and prevailed. Moses spoke directly with God on many occasions, and Moses wrote five books describing these experiences. Do you have trouble believing in God because he was superhuman? Could any human create an animal? We have animals. Hence some superhuman being had to exist, or we would have no animals. Do you know of any superhuman being, other than God, who could have created the animals? It takes no more blind faith to believe in God than it does to believe in George Washington. More will be said about the existence of God in the next Chapter.

Let's consider the basic features of our theory of creation. We will assume that God existed, and that he came down to earth from time to time during the past 540,000,000 years to create animals. As time went by, he made more complex animals. This simple theory accounts for all the animals, it accounts for the increase in complexity of the animals, and it accounts for the fossil record, including its sudden appearances of new animals, its gaps, its missing links, and all of the other characteristics of the fossil record. Since it fits the facts better, I think the theory of creation is more scientific than it is religious. And I think the theory of evolution is more religious than it is scientific.

In Conclusion

The affidavit of this Chapter presents the testimony of the evolutionists, and some analyses of this testimony. It is always appropriate for one expert witness to study and analyze the expected testimony of other witnesses. The next Chapter presents the affidavit of the Author of this book.

The Cambrian Explosion

CHAPTER 24. THE AFFIDAVIT OF YOUR AUTHOR.

Purpose of This Affidavit.

In connection with many of the law cases in which I have served as an expert witness, I was asked by the attorneys to prepare an affidavit, a sworn written statement recording some of my findings pertinent to the case. An affidavit, of course, is always prepared prior to the trial, and it may cover only part of the expert's testimony. The full testimony is presented at the trial. In Chapter 23 we analyzed affidavits offered by evolutionists, opinions and information contained in their writings, and some reasons for these opinions. This Chapter will serve as my affidavit on the subject of the question, "Does God really exist?" My full testimony will be given in the next Chapter.

The existence of God is pertinent to the subject of who designed the animals because, in the previous Chapters of this book, we have presented evidence which strongly suggests that some intelligent being had to be responsible for the design of the animals. And we have critically analyzed ancient existing written documents, specifically the books of the Bible, which claim that the person who designed the animals was the God of the Bible. Many evolutionists do not believe that God exists, or ever did exist, and as a result of this belief, they are compelled to be evolutionists. Therefore, whether or not God exists is an important issue germane to the subject matter of this book.

So far, in this book, I have accepted the challenge of finding out how the animals arose, both during the Cambrian Explosion, and since then. I have used my knowledge of engineering to demonstrate that animals are complex machines and they had to have been designed by somebody. We also studied several other areas of science and engineering and we showed that these analyses reinforced the above conclusion. We successfully used statistical probabilities, palentology, studies of various New-Feature-Producing Agents, an engineering analysis of the creation of the universe, the earth and its inhabitants. As a part of a responsible and scholarly investigation by an expert witness we then thought it was logical to ask who was it that did this designing, and who was it that caused all of this creation. Only then did we consider the possibility that the God of the Bible may have had something to do with the design and construction of the animals. We have been forced, as a result of engineering logic, and by the pursuit of facts, to consider the possibility that the God of the Bible designed the animals. If God exists, then that is a fact with which we must deal. It is not a matter of religious belief. And if we quote from the Bible, that is for the purpose of establishing facts, not to demonstrate our faith, conform to any doctrine, or to promote any particular religion.

Three Reasons Why I Believe that God Exists.

I will now present the substance of my affidavit. I do believe that God exists! There are many reasons for this, but I will discuss only three of them. My studies and my experiences have convinced me of the following: (1) The living plants and animals of the earth strongly testify to the existence of God; they are miraculous and supernatural. (2) The Bible, which reveals God, is a miraculous and supernatural Book, especially its contents which prophesy future events, and (3) I have experienced several events in my life that I consider to be related to God, and they were miraculous and supernatural. The natural forces of the earth are listed in Chapter 9. They are governed by the laws of physics and chemistry. In this Chapter we will be talking about supernatural forces.

Life Testifies to the Existence of God.

The other Chapters of this book cover the first reason, that the living creatures of this earth testify to the existence of God. All of the living things about us, because of their superlative design, tell us that there is a God. We will include the contents of this book that describe some of our animals, by reference, as a part of this affidavit, and we will not need to dwell further on that subject here.

The Prophesies of the Bible Testify to the Existence of God.

Secondly, the prophesies of the Bible, being miraculous and supernatural, testify to the existence of God, who claims to have written the Bible, by dictation to his secretaries, the various authors of the books of the Bible. I will identify just a few examples of these prophecies, using the King James Version of the Bible.

The Birth of Jesus Christ.
About 800 years BC, Isaiah wrote the following words:

> "Hear ye now, O house of David . . . the Lord himself shall give you a sign; Behold, a virgin shall conceive, and bear a son, and shall call his name Immanuel (which means 'God with us')" Isaiah 7:13, 14. And Isaiah also wrote, "the people . . . have seen a great light . . . For unto us a child is born, unto us a son is given, . . . and his name shall be called Wonderful, Counselor, The Mighty God, The Everlasting Father, the Prince of Peace." Isaiah 9:2, 6.

About 800 BC, Micah wrote:

> "But thou, Bethlehem . . . though thou be little among the thousands of Judah, yet out of thee shall he come forth unto me that is to be ruler in Israel; whose goings forth have been from old, from everlasting." Micah 5:2.

These are only three of the many verses in the Old Testament of the Bible which prophesy concerning the birth of Jesus Christ. About 800 years after these words were written, Jesus Christ was born, and he was born in Bethlehem. Could you predict the birth of some important person 800 years in the future, and foretell the town in which that person would be born?

Messiah the Prince.
About 550 BC Daniel wrote the following words:

> "Know, therefore, and understand, that from the going forth of the commandment to restore and to build Jerusalem unto Messiah, the Prince, shall be seven weeks, and threescore and two weeks . . ." Daniel 9:25.

This identification of a period of time separates two events. The first event was "the commandment to restore and to build Jerusalem." The only such commandment was given by the King, Artaxerxes, as recorded by Nehemiah about 450 BC, which was:

> "And it came to pass in the month of Nisan, in the twentieth year of Artaxerxes, . . . I (Nehemiah) said unto the king . . . wouldest send me unto Judah, unto the city of my fathers' sepulchers, that I may build it . . . So it pleased the king to send me." Nehemiah 2:1-6.

According to competent authorities, the time of this conversation, and the time of the King's commandment, was March 14 in the year 445 BC.

Now let's see if we can determine the date of the end of this time period. About 500 BC Zechariah wrote the following:

> "Rejoice greatly, O daughter of Zion, shout, O daughter of Jerusalem; behold thy King cometh unto thee; he is just, and having salvation; lowly, and riding upon an ass, and upon a colt, the foal of an ass." Zechariah 9:9.

About 500 years after Zechariah wrote this, Matthew and John recorded the following:

> "Then Jesus sent two disciples, saying unto them, Go into the village . . . and find an ass tied, and a colt with her, loose them and bring them unto me . . . and the disciples . . . brought the ass, and the colt, and they set him thereon . . . many people . . . took branches from palm trees, and went forth to meet him, and cried, Hosanna! Blessed is the King . . ." Matthew 21:1-7 and John 12:12-15.

We celebrate this event even today. It is called Palm Sunday. Fortunately, the exact day on which this event occurred can be established. Referring to this event, John wrote:

> "the Jews' Passover was nigh at hand." John 11:55.

John then wrote:

"Then Jesus, six days before the Passover, came to
Bethany . . . There they made him a supper, and Martha
served . . . On the next day . . . Jesus found a young ass,
sat upon it . . ." John 12:1-2.

At that time the Passover started on Nisan 14. Six days before that was
Nisan 8, which was Friday. The Sabbath would then have been Nisan 9,
the day of Martha's supper. The next day, the real Palm Sunday, was then
Nisan 10. This can be equated to April 6 of the year 32 AD.

Now let's calculate the amount of time involved in the original
prophecy of Daniel 9:25. In Biblical prophesies, seven years is often
referred to as one week, and 360 days is referred to as one year. For
example, Daniel calls 3.5 years, "a time and times and the dividing
of times," Daniel 7:25; or, "a time, times, and an half," Daniel 12:7.
Referring to the same 3.5-year time period, John says, "and the holy city
shall they tread under foot forty and two months," Revelation 11:2; and,
"the woman fled into the wilderness . . . that they should feed her there
a thousand two hundred and threescore days." Revelation 12:6.

Thus: 3.5 years = 42 months = 1260 days, and,
             one year = (1260/3.5) = 360 days.

Moses called a seven-week period one week. Genesis 25:27.

Now, using these commonly accepted Biblical number customs, the
time period of Daniel 9:25, starting with King Artaxerxes' command that
Nehemiah should rebuild Jerusalem, and extending to the triumphant
entry of Jesus into Jerusalem, on the colt of an ass, we get the following
predicted time,

$$T = (7 + 60 + 2)(7)(360) = 173,880 \text{ days.}$$

Now let's see if we can calculate the actual number of days that
transpired between March 14 of the year 445 BC, and April 6 of the year
32 AD. For this calculation we should use 365-day years, and we must
add or subtract days for leap years. The following calculations give us
the actual days between these two dates.

Days From Artaxerxes' Commandment to Messiah the Prince

Items that Must be Considered		Period of Time
Gross number of years	445 + 32	+477 years
Correction since from 1 BC to 1 AD is one year, not two		-1 year
Subtotal		+476 years
Gross number of days	(476)(365)	+17,740 days
Days from March 14 to April 6 (Including both days)		+24 days
Addition for normal leap years	476/4	+119 days
Subtraction for 100-year no-leap-year years		-3 day
Total Days		+173,880 days

Thus, we can conclude, that, if these calculations are accurate, Jesus Christ, Messiah the Prince, Messiah the future King, came riding into Jerusalem, on the colt of an ass, on the very day that was predicted by the prophet Daniel, more than 500 years before. Would you agree that only God could have made such a prediction?

Prophesies Relative to Israel.

In many places in the Bible it states that the descendants of Israel, whom we now call Jews, would be defeated in battle and would be scattered all over the world. It states that the Jews would be severely persecuted in these nations to which they had been scattered, and it then states that, in the latter days, the Jews would be reassembled in their old land, and the nation of Israel would be reestablished. All of these predictions have come to pass.

About 1450 BC Moses wrote, relative to the Israeli people:

> "But it shall come to pass, if thou wilt not hearken unto the voice of the Lord . . . the Lord shall cause thee to be smitten before thine enemies . . . and shalt be removed into all the kingdoms of the earth" Deuteronomy 28:15, 25.

About 750 BC Amos and Hosea wrote:

> "I will sift the house of Israel among all nations, as corn is sifted in a sieve," Amos 9:9.

> "My God will cast them away, because they did not hearken unto him; and they shall be wanderers among the nations" Hosea 9:17.

The heyday of the nation Israel was during the reign of King David, about 950 BC. But the above prophesies began to be fulfilled in 586 BC with the fall of Judah and the Babylonian captivity. Then, in 70 AD the Roman Empire finished the dispersion of the Jews. And there, among the nations of the world, the Jews have been for the past nineteen-hundred years.

The Bible then prophesied that the Jews would be downtrodden, oppressed and persecuted in the nations to which they had been scattered. About 1450 BC Moses wrote:

> "among these nations shalt thou find no ease, neither shall the sole of thy foot have rest; but the Lord shall give thee there a trembling heart . . . And thy life shall hang in doubt . . . and shalt have no assurance of thy life." Deuteronomy 28:65-66.

About 650 BC Jeremiah wrote:

> "Nevertheless, in those days, saith the Lord, I will not make a full end with you." Jeremiah 5:18.

About 550 BC Ezekiel wrote:

> "yet will I leave a remnant, that ye may have some that shall escape the sword among the nations, when ye shall be scattered through the countries" Ezekiel 6:8.

These prophesies have been remarkably fulfilled. In 70 AD the Romans destroyed Jerusalem, the Jewish Temple and the entire nation. Thousands of Jews were killed and those who survived were sent to the slave markets in Egypt. For almost two-thousand years the Jews have wandered around the world with no country of their own. During the Crusades, Jews were massacred. Laws were passed to prevent Jews from owning land or becoming farmers. They were forced to live in ghettos. Millions of Jews were murdered by Hitler's henchmen. But yet we still have a remnant of Jewish people with us today.

Finally, the Bible predicts that "in the last days", which many believe refers to the present, the Jews will return to their homeland, and establish again their nation, and rebuild the waste lands. About 750 BC Micah wrote:

> "But in the last days . . . saith the Lord will I assemble her that halteth, and I will gather her that is driven out, and her that I have afflicted; and I will make her . . . a strong nation" Micah 4:1, 6, 7.

In about 650 BC Jeremiah wrote:

> "And I will gather the remnant of my flock out of all countries to which I have driven them, and will bring them again to their folds; and they shall be fruitful and increase." Jeremiah 23:3.

We who are living today are witnessing this regathering of the scattered Jewish people. In the late 1800's and early 1900's the Jews began to return to their old homeland in Palestine. On May 14, 1948 Israel was again established as a nation. At that time there were 650,000 Jews in the new Nation. Since then the Jews have flooded back to their homeland. By 1984 the population of Israel was 4,200,000. It continues to grow. We are witnessing the fulfillment of these prophesies right before our eyes.

Is Israel important in the world today? Israel is only slightly larger in area than the State of New Jersey; but the political stability of the world today revolves around the affairs of Israel. In almost every newspaper every day, there are articles telling us what is going on in this tiny Nation.

There are also Biblical prophesies that foretell the future of Israel, but I will let you search in the Bible for them on your own, because that is not the primary subject of this book. However, there is one verse in the

Bible that greatly intrigues me, which I will share with you. The Bible predicts that the current world, as we know it, will eventually come to an end, and a period of 1000 years will begin which will be quite different. This is referred to by some as the millennium. The Jews will be involved in this new era. Would you like to know when this new 1000-year period will begin? I can't tell you, but the Bible contains one verse that might give you food for thought on this subject. First, recall that the Bible says that "one day is with the Lord as a thousand years." 2 Peter 3:8. Now read Hosea 6:2. It says:

> "After two days will he revive us; in the third day he
> will raise us up, and we shall live in his sight." Hosea 6:2.

Obviously, the "he" refers to God, and the "us" refers to the Jews. And each day is a thousand years. Israel was scattered among the nations about 2000 years ago. Israel is now being "revived". Are we near the time when "in the third day", the next thousand years, Israel will be raised up and will live in his sight? Other verses of the Bible explain in more detail what is expected to happen during these three thousand years. Open your Bible and read Hosea 6:2 for yourself, and tell me what you think! We seem to be near the time that the "two days" end and the "third day" is about to begin.

The prophesies concerning Israel cannot be explained away. I don't know of any human being who could make such prophesies. I have to conclude that God exists and that God wrote the Bible.

<u>Prophesies Relative to the Death of Jesus Christ.</u>

About 950 years before it happened, David, King of Israel, described the death of Jesus Christ in great detail. These seem to be the words and thoughts of Jesus Christ as he was being crucified.

> "My God, my God, why hast thou forsaken me? . . .
> All they who see me laugh . . . saying He trusted on
> the Lord . . . let him deliver him . . . I am poured out
> like water, and all my bones are out of joint: my heart
> is like wax; it is melted in the midst of my bowels . . .
> My strength is dried up . . . my tongue cleaveth to my
> jaws . . . they pierced my hands and my feet. I may tell all

> my bones . . . they parted my garments among them, and
> cast lots upon my vesture . . ." Psalms 22.

Isaiah wrote the following words 750 years before Jesus Christ was crucified:

> "He is despised and rejected of men . . . and we esteemed
> him not . . . smitten of God, and afflicted . . . and the Lord
> hath laid on him the iniquity of us all . . . yet he opened not
> his mouth, he is brought as a lamb to the slaughter . . . he
> made his grave with the wicked, and with the rich in his
> death . . . Therefore will I divide him a portion with the
> great . . . because he hath poured out his soul unto death . . .
> and made intercession for the transgressors." Isaiah 53.

About 550 BC Zechariah wrote the following words, which obviously refer to the fact that the disciple, Judas, betrayed his master for 30 pieces of silver. But Judas returned the money, and it was spent to buy a potter's field, and Judas hanged himself.

> "And I said unto them . . . give me my price . . . So
> they weighed for my price thirty pieces of silver, and cast
> them to the potter." Zechariah 11:12-13.

These prophesies, and about 300 more of them, which we will not have space to discuss, were made hundreds of years before they were fulfilled. But every one of them was fulfilled in every detail.

Prophesies Concerning World Empires.

Hundreds of years before the following events took place, the Bible, especially in the Book of Daniel, predicted the Babylonian Empire of 606 BC, the Media-Persian Empire of 530 BC, the Greek Empire of 331 BC, and the Roman Empire of 68 BC. The Bible also predicts that a great future battle will take place, against Israel, which will involve most of the countries of the world.

Conclusions Concerning Prophesies.

Hundreds of other examples of accurate predictions of future events could be cited, but the above examples should suffice to give you a good introduction to this subject. The prophesies of the Bible are supernatural. And the only supernatural being I know, is the God of the Bible. A study of such predictions has convinced me that God exists, and that God wrote the Bible!

Author's Personal Experiences Testify to the Existence of God

Thirdly, consider the following personal experiences which have happened to me during my lifetime.

<u>Does God Provide for the Physical Safety of Believers?</u>

When I was a teenager, we lived in LaGrange, Illinois, a suburb of Chicago. My cousin was a navy recruit undergoing basic training at the naval Base north of Chicago. He visited us one weekend, and, at the end of the visit, my parents asked me to drive him, in our family car, to the elevated-train terminal near Westchester. It was Sunday afternoon and he needed to return to his Base. As teenagers tend to do, I drove the large Buick at an excessive velocity as we sped north approaching Cermak Avenue. When we got near the busy east-west road I stepped with great force on the brake pedal, so we could get the speeding car stopped in time to maneuver safely onto Cermak Avenue. Much to my horror, the brake pedal slammed to the floor without producing any braking force whatever, and we continued to speed forward toward the busy highway. It was Sunday afternoon and Cermak Avenue was filled with closely-spaced speeding cars going in both directions. It was apparent that there was no way we could avoid colliding with one or more of the cars. But there was no collision! Like magic, somehow we slipped between two cars going east, and two other cars going west. We were untouched. Our Buick jumped a small ditch on the other side of the road and we then plowed forward into a vacant muddy field until the car stopped. With some help we got the Buick back on the road, and, by using the emergency brake, I succeeded in completing my mission and getting the car back to LaGrange. A pin had sheared in the mechanical brake linkage.

That evening, as was my custom, I went to Church, and, much to my amazement, the preacher's sermon topic for the evening was the contention that God provides for the personal safety of his believers. He called it the providential provision of safety, and he cited the following verses from the Bible.

> "Because thou hast made the Lord . . . thy habitation, There shall no evil befall thee . . . For he shall give his angels charge over thee, to keep thee . . ." Psalms 91:9-11.

> "The angel of the Lord encampeth round about those who fear him, and delivereth them." Psalms 34:7.

> "Are not two sparrows sold for a farthing? And one of them shall not fall on the ground without your Father. But the very hairs of your head are all numbered. Fear not, therefore; ye are of more value than many sparrows." Matthew 10:29-31.

I had just become a Believer shortly before this near-accident occurred, and I had been obsessed with the matter of learning whether or not God really existed, and whether or not my newly-adopted beliefs were based on facts or on myths. I zealously prayed to God requesting that, if he really existed, would be somehow make himself known to me. I then experienced this incredulous escape from death in the Buick, and I marveled at the remarkable coincidence that the preacher, that same night covered the topic of providential preservation. I was then convinced that these events were an answer to my prayer, and my belief that God existed was greatly reinforced.

Why Didn't the Snake Bite Me?

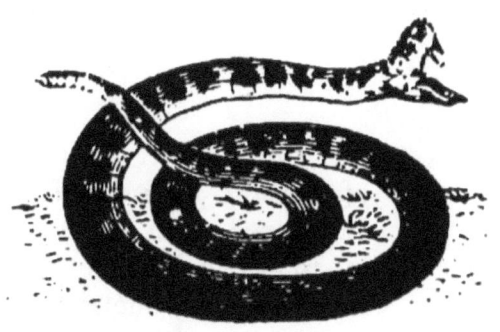

Some years later my wife, Bonna, and I were cruising in our boat from south Florida to Mobile, Alabama, and we had arrived at Apalachicola, Florida at the time we needed to stop and work on our income tax. The marina at Apalachicola had been demolished by a hurricane, but we found a live 110-volt electrical outlet in the wreckage, so we docked there for several weeks. After the tax work was completed we decided to stay one more day. We had noted on the chart that, about ten miles off shore, in the Gulf of Mexico, there was an Island that seemed to have a dock attached to it. So, for adventure, we cruised to the Island and tied up to the dock. We found that it was a deserted island, but there was an obsolete lighthouse on it, so we hiked to the lighthouse. On the way back to the boat we were walking along the trail that lead from the lighthouse to the dock. I was walking about 15 feet ahead of Bonna, and I was looking out to sea, not paying much attention to the trail. Suddenly, behind me I heard my wife scream. I looked back and observed that right in front of her, and where I had just passed, there was a large water-moccasin snake, right in the middle of the trail. Its head was up about 10" above the ground, in striking position, with its large mouth wide open, at an angle of about 120 degrees. I could see its teeth and the white cotton-mouth coloring inside of its mouth. It was about 5 feet long and slightly under 3 inches in diameter. I must have walked just a few inches in front of the ready-to-strike snake. Why it didn't strike me, I will never know. Bonna would have stepped right on the snake if she had not been looking at the trail. She saw the snake, and stopped.

If the snake had bitten me, death would have been the result. We were far out of radio range of any Coast Guard Station, and there were no other boaters monitoring the radio in that remote area of the Gulf. In fact, around the big bend area of Florida it is wilderness. We traveled 192 miles and did not see another cruising boat. And, Bonna had not yet learned how to pilot our boat. If the snake had bitten me on that deserted Island, I surely would have died. Is it possible that God spared my life? I remembered that the preacher had quoted the following: "His angels . . . will bear you up . . . lest you strike your foot against a . . . (serpent)".

Why Didn't the Jeep Roll Over the Cliff?

One summer a group of several of my friends had joined us in an attempt to travel in 4-wheel-drive Jeeps over a mountain near Crested Butte, Colorado. My son David was among the group. The road had been blasted out of the mountain about 100 years ago. It was largely a pile of rocks, and it was almost impassable, even for Jeeps. The road consisted of a narrow ledge cut out of the mountain, with a cliff on the downhill side which extended precipitously downward to several hundred feet below. One by one, the other members of the group dropped out of the challenge until, finally, my CJ-5 was the only one still moving upward. My wife and my daughter-in-law, Vicki, were with me.

We finally got to the top of the mountain and started down the other side. Shortly thereafter, we encountered a large snow drift in the road. I carefully examined the snow drift and decided that I could go through it. The road was slightly down hill, and that helped. We made it through the drift, but soon encountered an even larger snow drift which obviously was impossible to penetrate. So, we turned the Jeep around on the narrow ledge, using many back-and-forth excursions, turning the Jeep a few degrees during each sally.

We then needed to go back through the snow drift that we had earlier traversed. But this time it was different, because now we had to go uphill. I knew that going uphill through snow would likely cause the wheels to slip, and if they slipped, the Jeep would slide sidewise toward the cliff. Also, the snow bank was not level side-to-side. It sloped slightly downward toward the precipice. Other than trying to get through the snow drift, our only other options were to spend the night there, for which we were not prepared, or to walk 20 miles back to civilization.

I decided to try to drive through it. Bonna and Vicki got out of the Jeep. I selected the best gear and gave it the amount of gas that I thought would be best. I started through the snow drift, but then, my worst fear materialized. The Jeep wheels started to slip, and the Jeep began to skid sidewise toward the cliff. Then the left wheels went over the cliff! I looked down into the precipice and I thought I was history. Instinctively, I floored the accelerator and turned the wheels slightly more to the right. Bonna and Vicki looked on with horror. The Jeep continued to creep forward with two wheels over the cliff. Then I began to feel the front wheel grab the rocks at the edge of the cliff. Rocks flew in all directions, as I continued to floor the gas pedal. Finally, the left front wheel came slowly up onto the road, and then the left rear wheel followed. I was on the road and past the snow drift. Bonna and Vicki were crying. I had just escaped death!

Do you think that God helped me? Bonna and Vicki and I think he did. I remembered that the preacher had quoted the following: "For He will give His Angels charge concerning you, to guard you . . . they will bear you up in their hands."

He Who Lacks Wisdom, Let Him Ask of God.

When I was a sophomore in the College of Engineering at the University of Louisville, I had an English professor who was an atheist and a communist. He tried to recruit me to become a member of the Communist Party, and to be an atheist. He and I had many discussions, most of which were very demoralizing to me. Based on his lectures and his assignment of reading materials, it was obvious to me that his primary interests were atheism and communism, not English. Since he was older and better educated than I, obviously I was influenced by him, and I was devastated by the thought that maybe he was right and I was wrong. Possibly all of my beliefs and moral standards were based on myths and error.

After one long talk session with this Professor I was particularly confused and disturbed. He had succeeded in planting the seeds of doubt in my mind concerning the very existence of God. I was devastated. I went home and prayed fervently to God, asking him somehow to tell me whether or not he existed. Just then, I saw a copy of the Bible laying on the mantle over the fireplace. I walked over there and opened it, hoping that there might be some message in it which would be an answer to my prayer. When the Bible became open, my eyes immediately fell on these words:

> "My brethren, count it all joy when ye fall into divers temptations; Knowing this, that the trying of your faith worketh patience . . . If any of you lack wisdom, let him ask of God . . . and it shall be given him." James 1:2-5.

You can search the Bible from cover to cover and you won't find any other words that compare with these as a profound answer to my prayers. The Bible contains about 800,000 words. What do you think is the statistical probability that, by chance, I could have stumbled onto these particular 39 words? This obviously was the answer to my prayers. In fact, I considered it then, and I consider it now, to be a direct response from God to me. I consider it to be a miracle. And it proved to me that God exists.

Would God Bring About an Auto Accident to Help Me Select a Wife?

Throughout my youth I searched diligently for a wife. My standards were very high and I found it difficult to find a young lady that I thought met all of my exacting qualifications. Finally, I found one who, I thought, might make a fine wife, but I had just met her, and soon after we met she graduated with her MSc degree from Ohio State. Her parents came to attend the graduation ceremonies, and then they all got in her parents' car and headed for Florida. I assumed that I would never see her again. Then, an hour or so after they had departed, my phone rang. It was Bonna. She said they had been in an auto accident, and the trip to Florida for her would have to be canceled. Her parents then went by bus to Indiana where they had left their travel trailer, and I invited Bonna to go with me to my parents' home in Louisville for the Christmas holidays. There we had an opportunity to get very well acquainted. Bonna proved to be a phenomenally well qualified young lady, and well suited to me. A year later we were married, and we have had a most excellent, happy, idyllic and exemplary marriage, even to this day.

Did God cause that auto accident to happen, which brought Bonna and me together? You judge for yourself. I think he did. Does God exist? I think he does!

Would God Help Me to Write This Book?

After I retired from being a professor, and from serving as an expert witness in law cases, I felt a strong urge to write this book. Was God asking me to do it? I certainly could not prove that to you, but I can tell you of several incidents relative to the writing of this book which suggest to me that God approved of this project. Here is one incident.

Chapter 12 of this book explains how, by the flapping of their wings, birds can generate substantial aerodynamic lift. I also told you in Chapter 12 that I learned, directly from a bird, how this lift is generated. What I didn't tell you in that Chapter is that, after failing to learn how birds' wings work from aeronautical engineers and biologists, and from my own analyses, and since I wanted to write about birds in this book, I was highly motivated to find out how birds fly. So, one night I prayed to God asking him to show me how birds fly. The next day, while taking my usual walk, a dead bird fell out of a tree right in front of me, and, from this bird, I learned how a bird can fly by flapping its wings. The bird was warm and flexible. Figure 12.1 of Chapter 12 shows a picture of this bird. I could see the off-center shafts of its feathers, and I could clearly see how each wing acted as a one-way valve. I could blow on the birds' wings and watch these one-way valves work. I had no idea that off-center feathers were crucial to a bird's flight, until this bird clearly showed me.

Have you ever had a bird fall out of a tree right in front of you? I hope you appreciate that this was truly a miracle, a supernatural event! It was directly an answer to my prayer. And it proved again to me that God exists, and that he was encouraging me to continue to write this book.

Another book-related incident may be of interest. Bonna and I were driving to Colorado to go camping with our youngest son, John, and his family. He is a professor at Purdue. While driving, as usual, I was mentally engaged in doing planning relative to this book. I was planning Chapter 13, and I wanted to introduce the substance of that Chapter with an anecdotal tale about an anthill on the Oregon Trail. I kept asking Bonna what she knew about the Oregon Trail. She knew as little as I did but she found parts of it on the map. Then, when we finally met my son John and his family we discovered that, for reading entertainment, just before embarking on this trip, they had taken out a book from the library on the subject of, guess what, the Oregon Trail. I was then able to proceed with the writing of Chapter 13. Was that a coincidence, or is it possible that God led them to take out that book for my benefit.

After I had finished writing Chapter 9, on "Who Built the Fire Ring?", I was somewhat confused, not knowing what to write next. Again, I prayed to God asking him to tell me what I should write. Immediately thereafter, while leafing through a book on biology, I was startled as my eyes were drawn to a picture that looked like a fire ring similar to Figure 9.1 of Chapter 9, which I had just completed. It amazed me that there would be a picture of a fire ring in a biology book. Actually, it was a cross-section through the flagellum of a protozoan, similar to Figure 10.3 of Chapter 10. Thus, Chapter 10, on "Who Designed the Protozoa?" was initiated. When I decided to write about a fire ring, I had no idea that my fire ring would introduce the subject of a protozoan. Was this incident a mere coincidence? Maybe; but when such events occur repeatedly, it suggests to me that some of them might be engineered by God.

Who Caused Noah's Flood?

You have probably already concluded from the above accounts of my affidavit experiences that I often pray to God and ask him to reveal to me some matter of knowledge or wisdom. As a final example of such experiences, I am now going to share with you what I consider to be a most remarkable answer to one of my solicitations.

I have often wondered about the details of the flood described in the Bible, in which Noah and his family were preserved. My mind particularly wandered to this subject as I was engaged in some mental analyses related to the origin of the earth, which are presented in Chapters 20 and 21. One night, while in bed, I simply prayed to God and asked him to tell me how he arranged for it to rain for forty days and forty nights, and cause the great flood. Instantly, the following scenario came into my mind. None of the concepts described below had ever entered my mind before.

You should recall from the accounts presented in Chapter 20 that there was a time in the early history of the earth when there was a lot of water up in the atmosphere high above the earth, and this effectively shut out the sun's rays from reaching the earth. At that time the earth was heated by the molten rock under the earth's crust, not by the sun. Then, as the earth cooled, much of the water above the earth fell to the earth's surface where it filled the ocean, as described in Chapter 20. Then, the sun's rays could shine down upon the earth and the weather on the earth eventually became dominated by the sun rather than by the heat of the molten rock within the earth. When the heating of the earth became dominated by the sun, the earth would be very warm near the equator, but very cold at the north and south poles. There would then be a time period during which there would be a lot of moisture in the air and it would be colder than 32 degrees near the poles.

The result of this situation would be the build-up of a tremendous quantity of snow and ice at both poles. Such accumulated ice caps are known to have developed on Mars, and, I believe, on some of the moons of other planets. All that would be necessary then, to produce 40 days and 40 nights of rain would be for a few volcanic eruptions to take place on the earth at one or both of the poles. It is also known that in the past there was more volcanic activity on earth than in more recent times. And violent eruptions would particularly be expected to occur at the poles, after the build-up of the polar ice caps, because the earth, there, would have been overladen with the huge weight of these ice caps. The heat from the eruptions would melt and vaporize the accumulated ice, and this water vapor would then overflow through out the earth and fall as rain.

Of course God presumably could have made it rain by some other direct action, but it has been my observation that God often elects to accomplish his purposes using the natural forces of the earth as much as possible. It seems remarkable, indeed, to me, that this scenario of how God caused the flood to occur came instantly into my mind, just after I prayed to God and asked him to tell me how he did it. Of course you are welcome to scoff at the above account if you wish, but I would not be surprised to learn, eventually, that this revelation is an accurate narrative of just what happened when it rained for 40 days and 40 nights, as recorded in the Bible.

Summary of Affidavit

This affidavit is my sworn statement, in writing, on the subject of my opinions relative to the existence of God, including reasons for my opinions. This affidavit contains three parts.

First, the living plants and animals of the earth testify to the existence of God, the supernatural designer. The almost infinite complexity of the animals compels the fair-minded scholar to conclude that there had to have been a designer. The other Chapters of this book are included in this affidavit by reference, since they clearly speak to this subject.

Secondly, the prophesies of the Bible testify to the existence of God, because they appear to be supernatural. Only God could predict the future accurately and in detail, several hundred years before the events took place.

Thirdly, I have testified concerning several incidents in my life, which can only be described as miraculous and supernatural, which were obvious answers to my petitions to God, or his efforts to instruct me or provide guidance in making important decisions. In my opinion, the statistical probability that these many personal experiences could be merely coincidences is so low that I am forced to conclude that God does exist.

And, of supreme importance to the subject matter of this book is the concept that, if God exists, he could have been the supernatural engineer who designed and constructed the animals.

This affidavit covers only the results of my experiences and observations which relate to the question of whether or not God exists. My complete testimony on who caused the Cambrian Explosion will be presented in the next Chapter.

CHAPTER 25. WHO CAUSED
THE CAMBRIAN EXPLOSION?

The Great Mud Slide.

Figure 25.1 Shale in the Mountains

Many millions of years ago, at a location which is now in western Canada, at the edge of a sea, there was a high and steep earthen bank that overlooked the sea. About 540,000,000 years ago, during the Cambrian period, many tons of this bank suddenly slid into the sea. It no doubt was a huge mud slide. It happened that at the bottom of the sea there were hundreds of marine animals, including creatures of a wide variety of types and sizes. These animals were trapped by the mud, and, due to its great weight, and its mineral composition, these animals, over the years, were turned into fossils. As the years went by, the mud turned into shale, the area uplifted, and, in 1910, at Burgess Pass in the Rocky Mountains, near Field, British Columbia, these fossil animals were discovered by human beings. Figure 25.1 shows shale such as at Burgess. The mud and silt in which the animals had been buried was so fine-grained that the animals were replicated in meticulous detail. These are now the several thousand fossils of the dark gray Burgess shale, which are securely locked in the cabinets on the 2nd floor of the Smithsonian Institution in Washington, DC. More recently, these same types of animals have been found in fossil beds in Greenland, China, Siberia and Namibia. All of the animals are of the same date, about 540 mya.

A Unique Group of Animal Fossils.

This group of animal fossils is unique, in all the history of the earth, for several reasons. First, they are very diverse in body plan and size. Figure 25.2 shows several of these animals. They were mostly marine invertebrates, but the November, 1995, issue of *Nature* reported the finding, in China, of a 2-inch-long primitive chordate, which, by its age, would be a member of this group. These animals included bristle-worms, lamp-shells, jellyfish, arthropods, ancient relatives of lobsters, crabs and spiders, trail-making and burrowing worms, starfish, etc. In fact the diversity of these animals was so great that they have been described as representing the whole animal kingdom. Scientist Conway Morris stated that in this group of animals, "nature invented the animal body plans that define . . . phyla." (Morris 1) Biologist, Cecie Starr, stated that "nearly all major animal phyla evolved during Cambrian times . . ." (Starr 8) Researcher, Michael Levine, has found that "Many new species have appeared since then. But nature drafted few, if any, new body plans after the Cambrian." (Levine 1) Biologist, Rudolf Raff, said "we've had these same old body plans for half a billion years." (Raff 1)

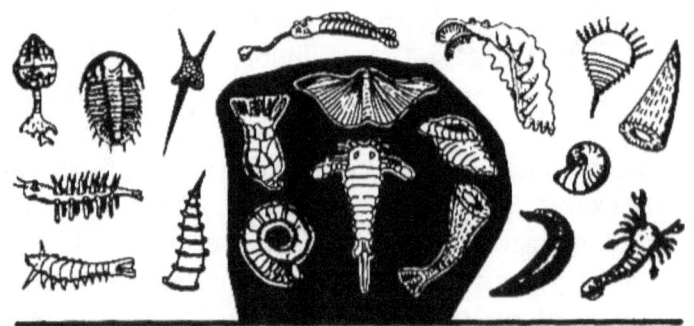

Figure 25.2. Animals From the Cambrian Explosion 543 mya.

Secondly, these animals were advanced in design and in features. One animal, Opabinia, had five large advanced eyes, and a proboscis that resembled a fire hose. A trilobite had some 20,000 eyes, a system apparently more advanced than that of any arthropod of today. (Luria 1) Another animal was a three-foot-long shrimp-like predator that grasped its prey in circular jaws, which, mechanically, acted like a camera shutter. Another looked like a legless lobster with an elephant's proboscis that ended in a pair of claws. These animals had bodies, limbs, and internal organs of great sophistication. Stephen Gould called them "Weird Wonders". Just look at the animals of Figure 25.2. Do they look like animals that have just evolved a step or two from a protozoan?

Thirdly, and by far the most sensational attribute of these Cambrian animals, is the fact that they all appeared suddenly, almost instantaneously, and without any ancestors. The sudden and instantaneous appearance of this large group of advanced and fully-developed animals has been likened by scientists to an explosion. Hence, the sudden arrival of these animals is widely referred to as the "Cambrian Explosion." The feature article in the *Time* magazine issue of December 4, 1995, was an account of the Cambrian Explosion. They called it "Evolution's Big Bang." The information contained on the cover page of this issue is shown in Figure 25.3. In this article, author, Madeleine Nash, wrote, "Then, 543 million years ago, in the early Cambrian . . . creatures with teeth and tentacles and claws and jaws materialized with the suddenness of apparitions. In a burst of creativity like nothing before nor since, nature appears to have sketched out the blueprints for virtually the whole of the animal kingdom." (Nash 1)

TIME

EVOLUTION'S BIG BANG

New discoveries show that life as we know it began in an amazing biological frenzy that changed the planet almost overnight
December 4, 1995

Figure 25.3 TIME cover information. Features the Cambrian Explosion

A Detailed History of the Cambrian Explosion.

When did this Explosion occur? Based on new studies in the 1990's, Nash wrote, "Virtually everyone agrees that the Cambrian started almost exactly 543 million years ago." (Nash 2)

We have used the word, "instantaneous" to describe the suddenness with which these animals appeared. How long did it take for this creative activity to be completed? Science editor Mark Hartwig reported that "discoveries in 1992 and 1993 have shrunk the explosion's estimated duration . . . to about 5 million years." (Hartwig 1) The earth is 4.6 billion years old. Five million years is 0.0001 (one one-hundredth of one percent) of the age of the earth. In geological time, this is, indeed, instantaneous.

279

When evolutionists find fossils of dozens of new, fully-formed, advanced and complex animals, they immediately begin to search for the ancestors of the new animals. They assume that they must have evolved from some predecessors through many millions of years of evolution. But in the case of these Cambrian-Explosion animals, their ancestors cannot be found. Mark Hartwig, stated that "in an instant of geological time, almost every animal phylum seemingly popped into existence from nowhere." (Hartwig 2) Schindewolf wrote, "In the Cambrian rocks, we encounter for the first time . . . an abundance of well-preserved and clearly interpretable fossils." (Schindewolf 7) Oxford evolutionary zoologist, Richard Dawkins, referring to the Cambrian-Explosion animals, stated, "It is as though they were just planted there, without any evolutionary history."

With respect to the assertion that the Cambrian-Explosion animals constituted the first large group of complex animals ever to inhabit the earth, let's direct our attention to the history of the earth prior to the Cambrian period. Paleontologist, Romer, wrote "Below this Cambrian period, there are vast thicknesses of sediments in which the progenitors of the Cambrian forms would be expected. But we do not find them . . ." (Romer 1) Luria, Gould and Singer wrote, "Geologists have discovered many unaltered Pre-Cambrian sediments, and they contain no fossils of complex organisms." (Luria 2) Zoologist Harold Coffin wrote, "If progressive evolution from simple to complex is correct, the ancestors of these full-blown living creatures in the Cambrian should be found; but they have not been found . . ." (Coffin 1)

To complete the history of the Cambrian Explosion we should ask whether or not there were any animals at all which lived prior to the Cambrian period. The answer is that there were some living creatures in the Pre-Cambrian eras. The earth is about 4.6 billion years old. During the period from 3.5 bya to 540 mya the rocks have shown that some single-celled life forms did exist, such as bacteria, algae, plankton and protozoa. Then, shortly prior to the Cambrian Explosion, a few fossils of other life forms appeared. Schindewolf reported that "in the younger sections of the Pre-Cambrian era . . . sparse remains of . . . sponge . . . worms, clams, snails . . . crustaceans . . . brachiopods . . . annelids and . . . arthropods have also been found." (Schindewolf 8) Morris has found that "the fossil record thus indicates that the Late Precambrian was dominated by . . . coelenterates that may have included all four living cnidarian (radially symmetrical) classes." (Morris 2) Gehling stated that "there is also evidence for the presence of arthropods (spiders, crustaceans, etc.) as well as echinoderms (sea stars, etc.) before the beginning of the Cambrian." (Gehling 1)

My interpretation of the significance of these findings is that whoever brought about the Cambrian Explosion did a little practicing with prototypes before proceeding with the production models. Look, again, at the vertical spikes that precede all of the subsequent expansions below these spikes, in Figure 22.7 of Chapter 22. These spikes, in my opinion, represent periods of experimentation with prototypes. It may also be true that some of the earlier single-celled plants and animals may have been employed to produce oxygen and soil. But the possible discovery of a few prototype animals prior to the Cambrian Explosion certainly does not in any way detract from the fact of the sudden and instantaneous appearance on earth of thousands of complex Cambrian-Explosion animals.

The Final Testimony of Your Author in Court.

We are now ready to hear the final concluding testimony of your Author. In accordance with the format of this book, you, the Reader, are the Judge and the Jury. The questioning Attorney will ask the questions, and I, the Expert Witness, will testify and reveal the findings from my investigations. In the following, "Q" stands for "Question", and "A" stands for "Answer". The court trial will now be in session. The Attorney who employed the Witness will do the questioning.

Q1: Dr. Starkey, you have been qualified by the Judge to be an expert witness in this trial. (See Chapters 2 and 3). What is your field of expertise?

A1: My expertise is in the field of machine design, which is a sub-field of mechanical engineering.

Q2: What was your assignment in this law case?

A2: I was asked to determine who caused the Cambrian Explosion, and to determine who designed and constructed the animals of the earth. (See Chapter 4)

Q3: Can you explain why your area of expertise, machine design, qualifies you to perform analyses in the field of the design of animals?

A3: An engineering analysis of animals clearly shows that animals are machines. They take in energy-producing fuel, the fuel is transformed into mechanical work, they have a frame and moving parts, they apply the principles of engineering design, and they are composed largely of machine elements. Since animals are machines, I am qualified to analyze them. (See Chapters 6, 7, and 8)

Q4: What has been your overall plan-of-attack to investigate the matter of who caused the Cambrian Explosion?

A4: I first had to determine if the animals were designed by: (1) somebody, or (2) nobody; and, secondly, if they were designed by somebody, then I needed to determine who was that somebody.

Q5: If the Cambrian Explosion was caused by nobody, and the animals were designed by nobody, how could these animals have come into existence?

A5: One of the theories of origins, the theory of evolution, purports to explain how the animals could have come into existence with nobody as a designer.

Q6: Would you please elaborate on the claims of the theory of evolution?

A6: According to the theory of evolution, somehow, the natural forces of the earth, without the assistance of any designer-craftsman, caused the sea-water to turn into a protozoan, and the protozoan, similarly, spontaneously evolved into animals, to produce the Cambrian Explosion, and to produce all of the animals which subsequently have inhabited the earth.

Q7: Have your investigations persuaded you to conclude that the theory of evolution is factual, and that the Cambrian Explosion was caused by nobody?

A7: No.

Q8: Would you explain why you have come to that conclusion?

A8: After concluding that any successful theory of evolution had to have two elements: (1) some effective New-Feature-Producing Agent (NFPA), and (2) natural selection, I studied all of the proposed NFPA's, including Mother Nature, teleology, adaptations, mutations, Mendelian genetics, punctuated equilibrium, and natural selection alone, and I concluded that none of these NFPA's could invent and design new features or produce new animals. (See Chapters 13, 14, 15, 16, 17, 18 and 19)

Q9: What about the fossils; didn't you find that they proved that the theory of evolution was true?

A9: No.

Q10: Please explain your studies of fossils.

A10: The fossil record shows that the animals of the Cambrian Explosion came into existence suddenly and almost instantaneously; and all of the other animals, similarly, from time to time over many years, came into existence suddenly, without any clearly-defined fossil trails of previous ancestors, half-finished body parts, transitional species, lineages, or family trees. The fossils do not support the theory of evolution. (See Chapter 22)

Q11: Have your investigations led you to the conclusion that somebody, some being, caused the Cambrian Explosion and designed the animals?

A11: Yes.

Q12: Please explain why you have concluded that somebody was the designer.

A12: I am an expert in the field of machine design. I can study a machine and clearly visualize the thought processes and the applications of the principles of design that went through the brain of the designer as he designed the machine. Animals are machines; and, similarly, I can visualize the thought processes of their designer, as he designed the animals. I have studied, organized and explained the various fingerprints of a designer, in such fields as materials-selection, form-synthesis of parts, mechanical, chemical and electrical systems, and the invention of clever and novel mechanisms. (See Chapter 5) Applying my understanding of the above fingerprints, I then studied many animals, including protozoa, spiders, birds, bees, bats, sex organs, the human arm, neurons, poisonous sea-creatures, squids, etc. And I studied such a simple construction as a 10-rock fire-ring on a mountain-top. From these analyses I concluded that none of the animals of the Cambrian Explosion, or any of the other subsequent animals of the earth, could have come into existence without the services of some real being as a designer. Based on my expertise in the field of machine design, I know that there had to have been a designer of the animals! (See Chapters 9, 10, 11, and 12)

Q13: Are you, at this point in our trial, testifying that you have arrived at the above conclusion with a high degree of engineering certainty? Are you sure that there had to have been *somebody* who designed the animals?

A13: Yes!

Q14: If somebody caused the Cambrian Explosion and designed and constructed the animals of the earth, have you determined the identity of that somebody?

A14: Yes.

Q15: Then who was it that caused the Cambrian Explosion?

A15: It was the God of the Bible.

Q16: And how do you know it was the God of the Bible?

A16: There are three reasons: (1) God is the only one who could have done it. (2) In a written document, God says he did it. And, (3) many human beings have testified, in writing, that he did it.

Q17: Would you elaborate on your answer?

A17: First, obviously, to design and construct an animal takes great creative capability. The only creatures we know who have any creative talent at all, are animals, humans, and God. We can say, categorically, that neither animals nor humans have enough creative capability to design animals. That leaves God as the only qualified candidate. Secondly, in the Bible, which is an ancient written document that claims to have been written by God through his secretaries, God says he created the animals. May I quote:

> "And God said, Let the waters bring forth abundantly the moving creature that hath life, and fowl that may fly above the earth in the open firmament of heaven. And God created great whales, and every living creature that moveth, which the waters brought forth abundantly, after their kind, and every winged fowl after his kind . . . And the evening and the morning were the fifth day.
>
> And God said, Let the earth bring forth the living creature after his kind, cattle, and creeping thing, and beast of the earth after his kind . . . And God made the beast of the earth after his kind, and cattle after their kind, and everything that creepeth upon the earth, after his kind . . . And God said, Let us make man in our image . . . And the evening and the morning were the sixth day." Genesis 1:20-31.

Thirdly, most of the "Authors" of the sixty-six books of the Bible have testified that God created the animals of the earth. Thus, God is the only one capable of creating animals, he said in writing that he did it, and many human beings have testified in writing that it was God who did it. (See Chapters 20 and 21)

Q18: Are we to understand that you believe that God exists? If so please elaborate.

A18: Yes. I believe that the existence of the universe, the earth, and the living things on the earth abundantly testify to the existence of God. Also, the miraculous prophesies of the Bible testify to his existence. And, some of my personal experiences have convinced me that God exists. (See Chapter 24)

Q19: Is it your testimony, then, that it was the God of the Bible who caused the Cambrian Explosion?

A19: Yes.

Q20: Are you testifying as an expert witness that you have arrived at the above conclusion with a high degree of engineering certainty? Are you sure that it was *God* who caused the Cambrian Explosion?

A20: Yes!

Q21: Can you elaborate at all on how God accomplished these acts of creation?

A21: I believe God created the universe and the earth, and then, after the earth cooled sufficiently, he came down to earth from time to time to design and construct the plants and animals. I believe the Cambrian Explosion of 540 mya was the fifth day of creation described in the Bible; (Genesis 1:20-23) and the following 540 million years occupied the sixth day of creation. (Genesis 1:24-31) The Cambrian Explosion involved marine animals. The fifth day of creation involved marine animals. The sixth day involved the other animals. The Cambrian Explosion was the beginning, the "Grand Opening", of God's creation of animals on the earth. He then continued that process for the next 540,000,000 years.

Q22: After coming to the above conclusions, what do you think of the theory of evolution?

A22: I think the theory of evolution is the greatest scientific mistake in the history of mankind. I love its proponents. I bear them no ill will, personally. But I think the theory they espouse is the greatest scientific mistake of all time!

Q23: Tell us, then, what is your overall conclusion regarding this matter.

A23: God exists! God created the universe! God recorded his deeds in the Bible! And God caused the Cambrian Explosion!

The Cambrian Explosion

BIBLIOGRAPHY

Alda 1: *Flight*, Narrated by Allen Alda, VHS Tape by Scientific American Frontiers.

Alvarez 1: *Science* 208, L.W. Alvarez, 1980, pp. 1095-1108.

Anderson 1: *Exploring the World Using Protozoa*, Anderson and Druger, 1997, p. 27.

Ayala 1: *Scientific American*, September, 1978, Francisco Ayala.

Bender 1: *Human Body*, Lionel Bender, 1992, p. 69.

Boucot 1: *Evolution and Extinction Rate Controls*, Arthur J. Boucot, 1975, p. 196.

Brackenbury 1: *Insects in Flight*, John Brackenbury.

Coffin 1: *Liberty*, Sept. 1975, Harold G. Coffin, p. 12.

Craig 1: *Prehistoric Facts*, Annabel Craig, 1986, p.31. All Craig references are reproduced from *The Usborne Book of Prehistoric Facts* by permission of Usborne Publishing, 83-85 Saffron Hill, London, ECIN 8RT, UK. Copyright 1987 Usborne Publishing Ltd.

Craig 2: *Ibid*, p. 45.

Craig 3: *Ibid*, p. 16.

Craig 4: *Ibid*, p. F.

Cromer 1: *The Miracle of Flight*, Richard Cromer, 1968.

Darwin 1: *The Origin of Species*, Charles Darwin, 1859, pp. 83, 88, 91, 92,

Dott 1: *Evolution of the Earth*, Robert H. Dott, Jr. and Roger L. Batten, 1988.

Dott 2: *Ibid, p. 148.*

Gehling 1: *Alcheringa* 11, 1987, J. G. Gehling, p. 337.

Geist 1: *Natural History*, Valerius Geist, 1981, p. 30.

Godfrey 1: *Scientists Confront Creationism*, Laurie R. Godfrey, 1983, p.158.

Godfrey 2: *Ibid*, p.158.

Godfrey 3: *Ibid*, p. 237.

Gould 1: *Paleobiography*, Winter 1980, Stephen Gould.

Gould 2: *Natural History*, May 1977, Stephen Gould, p. 14.

Gould 3: *Natural History*, Vol. LXXXVI(6), Stephen Gould, 1977, p.24.

Hartwig 1: *Moody*, Vol. 95, No. 9, 1955, Mark Hartwig, p. 16.

Hartwig 2: *Ibid*, p. 16.

Hickman 1: *Biology of Animals*, Hickman, Roberts and Hickman, pp.19, 21, Copyright 1986 by Times Mirror/Mosby College Publishing. All Hickman references are reproduced with permission of The McGraw-Hill Companies.
Hickman 2: *Ibid*, p. 463.
Hickman 3: *Ibid*, p. 257.
Hickman 4: *Ibid*, p. 24.
Hickman 5: *Ibid*, p. G-12.
Hickman 6: *Ibid*, p. 89.
Hickman 7: *Ibid*, p. 98.
Hickman 8: *Ibid*, p. 112.
Hickman 9: *Ibid, Back cover.*
Hickman 10: *Biology of Animals*, Hickman, Roberts and Hickman, 1986.
Hickman 11: *Ibid*, p. 109.
Hickman 12: *Ibid*, p. 109.
Hickman 13: *Ibid*, p. 598.
Hickman 14: *Ibid*, p. 13.
Hickman 15: *Ibid*, p. 237.
Hickman 16: *Ibid*, p. 433.
Hickman 17: *Ibid*, p. 368.
Jarvik 1: See Hickman, p. 549.
Keller 1: *Environmental Geology*, Edward A. Keller, 1988.
Kier 1: *New Scientist*, January 15, 1981, Porter Kier, p. 129.
Kitts 1: *Evolution*, vol. 28, 1974, David B. Kitts, p. 467.
Levine 1: *U. S. News*, Vol. 123, No. 7, M. Levine via S. Brownlee, p. 76.
Lubenow 1: *Bones of Contention*, Marvin L. Lubenow, 1992. All Lubenow references are reproduced with the permission of Baker Book House.
Lubenow 2: *Ibid*, p. 32.
Lubenow 3: *Ibid*, p. 171.
Lubenow 4: *Ibid*, pp. *73, 74*.
Luria 1: *A View of Life*, Luria, Gould and Singer, 1981, p. 638.
Luria 2: *Ibid*, p. 651.
Marshek 1: *Design of Machine and Structural Parts*, Kurt Marshek, 1987.
Morris 1: *Time*, Vol. 146, No. 23, 1995, S. Conway Morris, p.70.
Morris 2: *Nature*, Vol. 361, 1993, S. Conway Morris, p. 219.

Nash 1: *Time*, Vol. 146, No. 23, 1995, J. Madeleine Nash, p. 68. The Nash references are reproduced with the permission of TIME magazine.

Nash 2; *ibid*, p. 70.

Patterson 1: *Darwin's Enigma*, Luther Sunderland, 1984, p. 89.

Raff 1: *Time*, Vol. 146, No. 23, 1995, R. Raff via Madeleine Nash, p. 74.

Raup 1: *Field Museum of Natural History Bulletin*, January 1979, David M. Raup, p. 23.

Raup 2: *Ibid*, p. 22.

Ridley 1: *Evolution*, Mark Ridley, 1993, p. 5.

Ridley 2: *Ibid*, p.5.

Ridley 3: *Ibid*, p. *520*.

Ridley 4: *Ibid*, p. 601.

Romer 1: *Natural History*, Oct. 1959, Alfred Romer, p. 466.

Ross 1: *The Creator and The Cosmos*, Hugh Ross, 1993, p. 79.

Ross 2: *Ibid*, p. 118.

Ross 3: *Creation and Time*, Hugh Ross, 1994, p. 151.

Schindewolf 1: *Basic Questions in Paleontology*, Otto Schindewolf, pp. 311, 312, 343. Published by, and permissions granted by, the University of Chicago Press.

Schindewolf 2 *Ibid*, pp. 332-333.

Schindewolf 3: *Ibid*, p. 360.

Schindewolf 4: *Ibid*, pp. 168, 214.

Schindewolf 5: *Ibid*, pp. 186, 187.

Schindewolf 6: *Ibid*, p. 103.

Schindewolf 7: *Ibid, p. 16.*

Schindewolf 8: *Ibid*, p. 15.

Schindewolf 9: *Ibid*, pp. 215-216.

Schroeder 1: *Genesis and the Big Bang*, Gerald L. Schroeder, 1990.

Simpson 1: *The Major Features of Evolution*, George Gaylord Simpson, 1953, p. 360.

Snelling 1: *Quotebook*, Andrew Snelling, 1990, p.10.

Stanley 1: *The New Evolutionary Timetable*, Steven M. Stanley, 1982, p. XV.

Starr 1: *Biology Concepts and Applications*, Cecie Starr, 1991, p. 148, Copyright 1990. All Starr references are reproduced with the permission of Brooks-Cole Publishing Company, a division of International Thomson Publishing, Inc. All rights reserved.

Starr 2: *Ibid*, p. 179.

Starr 3: *Ibid*, p. 181.
Starr 4: *Ibid*, p. 180.
Starr 5: *Ibid*, p. 7.
Starr 6: *Ibid*, p. 179.
Starr 7: *Ibid*, p. 200.
Starr 8: *Ibid*, p. 202.
Stringer 1: *Natural History*, Christopher Stringer, 1984, p. 12.
Terres 1: *How Birds Fly*, John K. Terres, 1968.
Trinkaus 1: *Natural History*, Eric Trinkaus, 1978, p. 58.
Valentine 1: *Paleobiology*, Valentine, Collins and Meyer, 1994, p. 131.
Wagman 1: *Medical and Health Encyclopedia*, 1977, Vol. I,P. 41.

POSTSCRIPT

Based on the facts, engineering analyses and personal experiences described in this book, I came to the conclusion that the animals of the earth had to have been designed and constructed by someone, and I determined that this someone was, in fact, the God of the Bible. These conclusions were based on engineering reasoning, not on faith or religious doctrine. However, it should be appreciated that the events covered in this book were recorded primarily in the first chapter of the first book of the Bible. If you have read this book, and have come to agree with me that God exists, God created the animals and God wrote the Bible, then it would be logical for you to ask, "What does God say in the rest of the Bible? Of what significance is it to me if God created the animals? Does the rest of the Bible have any message for me? If so, what is that message, and what should be my response to that message?"

If you are interested in my answers to the above questions, please write to me at the address given below, and I will respond to you. I may even consider writing a book to answer these questions.

May the God of the Bible, who really exists and who lives today, richly bless you!

WALTER L. STARKEY
6503 COOK ROAD
POWELL, OHIO 43065

The Cambrian Explosion

INDEX

The letter *t* following a page number denotes a table; the letter *f* following a page number denotes a figure.

ORDER FORM

The book, The Cambrian Explosion, ID: 100448, exists in two
editions, namely, the 1999-edition and the 2012-edition.
These two editions are essentially identical except that the
1999-edition contains some pages of acknowledgments
and endorsements which are not in the 2012-edition.

If you wish to purchase a few copies of the 2012-edition
they should be available for sale in most book stores.

If you are interested is helping with the commercial
distribution of more than a few copies of the
2012-edition contact the following:
Faye.Cristobal@Xlibris.com
1-888-795-4271
Xlibris Corporation
1663 Liberty Drive, Suite 200
Bloomington, Indiana 47403

If you would like to purchase copies of the slightly less
expensive 1999-edition they can be ordered from:
Walter L. Starkey
614-323-3512
starkeywlsx@att.net.